Bodies, Affects, Politics

RGS-IBG Book Series

For further information about the series and a full list of published and forthcoming titles please visit www.rgsbookseries.com

Published

Bodies, Affects, Politics

The Clash of Bodily Regimes

Steve Pile

WILEY

This edition first published 2021

© 2021 Royal Geographical Society (with the Institute of British Geographers)

This Work is a co-publication between The Royal Geographical Society (with the Institute of British Geographers) and John Wiley & Sons Ltd.

The right of Steve Pile to be identified as the author of this work has been asserted in accordance with law.

Registered Office
John Wiley & Sons, Inc., 111 River Street, Hoboken, NJ 07030, USA
John Wiley & Sons Ltd, The Atrium, Southern Gate, Chichester, West Sussex, PO19 8SQ, UK

Editorial Office
9600 Garsington Road, Oxford, OX4 2DQ, UK

For details of our global editorial offices, customer services, and more information about Wiley products visit us at www.wiley.com.

Wiley also publishes its books in a variety of electronic formats and by print-on-demand. Some content that appears in standard print versions of this book may not be available in other formats.

Limit of Liability/Disclaimer of Warranty
The contents of this work are intended to further general scientific research, understanding, and discussion only and are not intended and should not be relied upon as recommending or promoting scientific method, diagnosis, or treatment by physicians for any particular patient. In view of ongoing research, equipment modifications, changes in governmental regulations, and the constant flow of information relating to the use of medicines, equipment, and devices, the reader is urged to review and evaluate the information provided in the package insert or instructions for each medicine, equipment, or device for, among other things, any changes in the instructions or indication of usage and for added warnings and precautions. While the publisher and authors have used their best efforts in preparing this work, they make no representations or warranties with respect to the accuracy or completeness of the contents of this work and specifically disclaim all warranties, including without limitation any implied warranties of merchantability or fitness for a particular purpose. No warranty may be created or extended by sales representatives, written sales materials or promotional statements for this work. The fact that an organization, website, or product is referred to in this work as a citation and/or potential source of further information does not mean that the publisher and authors endorse the information or services the organization, website, or product may provide or recommendations it may make. This work is sold with the understanding that the publisher is not engaged in rendering professional services. The advice and strategies contained herein may not be suitable for your situation. You should consult with a specialist where appropriate. Further, readers should be aware that websites listed in this work may have changed or disappeared between when this work was written and when it is read. Neither the publisher nor authors shall be liable for any loss of profit or any other commercial damages, including but not limited to special, incidental, consequential, or other damages.

Library of Congress Cataloging-in-Publication Data
Name: Pile, Steve, 1961– author.
Title: Bodies, affects, politics : the clash of bodily regimes / Steve Pile.
Description: Hoboken, NJ : Wiley, 2021. | Series: RGS-IBG book series |
    Includes bibliographical references and index.
Identifiers: LCCN 2020038677 (print) | LCCN 2020038678 (ebook) | ISBN
    9781118901984 (cloth) | ISBN 9781118901977 (paperback) | ISBN
    9781118901953 (adobe pdf) | ISBN 9781118901946 (epub)
Subjects: LCSH: Human body–Social aspects. | Human body–Political
    aspects. | Social distance–History–21st century. |
    Communities–History–21st century.
Classification: LCC HM636 .P48 2021 (print) | LCC HM636 (ebook) | DDC
    306.4–dc23
LC record available at https://lccn.loc.gov/2020038677
LC ebook record available at https://lccn.loc.gov/2020038678

Cover Design: Wiley
Cover Image: Justice for Grenfell protest (Saturday 16th June 2018) featuring Love4Grenfell logo
tee-shirt designed by Charlie Crockett and Cheshona Hart © Steve Pile

Set in 10/12pt Plantin by SPi Global, Pondicherry, India
Printed and bound by CPI Group (UK) Ltd, Croydon, CR0 4YY

The information, practices and views in this book are those of the author and do not necessarily reflect the opinion of the Royal Geographical Society (with IBG).

10 9 8 7 6 5 4 3 2 1

Contents

List of Figures

Series Editors' Preface

The RGS-IBG Book Series only publishes work of the highest international standing. Its emphasis is on distinctive new developments in human and physical geography, although it is also open to contributions from cognate disciplines whose interests overlap with those of geographers. The Series places strong emphasis on theoretically-informed and empirically-strong texts. Reflecting the vibrant and diverse theoretical and empirical agendas that characterize the contemporary discipline, contributions are expected to inform, challenge and stimulate the reader. Overall, the RGS-IBG Book Series seeks to promote scholarly publications that leave an intellectual mark and change the way readers think about particular issues, methods or theories.

For details on how to submit a proposal please visit:
www.rgsbookseries.com

Ruth Craggs, *King's College London, UK*
Chih Yuan Woon, *National University of Singapore*
RGS-IBG Book Series Editors

David Featherstone, *University of Glasgow, UK*
RGS-IBG Book Series Editor (2015–2019)

Preface

Today, London is quieter than I have ever known it to be. The skies above are undisturbed by the noise of planes, no white vapour trails scratching the brilliant blue. The East Coast mainline normally rumbles with heavy goods trains punctuated by the shattering sound of fast inter-city services, but not for the last two weeks. Normally, the day is interrupted by at least one low-level fly-over by a police helicopter, but not recently. The hum of traffic is notably subdued, as when snow falls, muffling sound, preventing vehicles from moving around the city. This quieting, however, is not a sign that the city is calmer, rested, at peace. Instead, the quiet feels more like frustration, determination and a low-level anxiety that threatens to break cover.

As I wait patiently in the queue at my local supermarket, I am paying attention to who is – and who is not – wearing face coverings, but especially noting the facemasks. Facemasks are as sure a measure of the level of anxiety and fear in the city as the intensification of the policing of bodies (which is not only conducted by the police). I know I am 2 metres back from the person in front of me and that the person behind me is 2 metres away from me. I know because the pavement has suddenly become covered in sticky tape that tells bodies where they should be. Sometimes, there are big stickers with footprints; 'stand here' they instruct. I am self-policing. I stand where I should, as do most people. Some people do not. They are policed: the supermarket has employed a company that, judging by their jackets, normally stewards entertainment events. A woman in a high vis jacket, continually adjusting her ill-fitting facemask, waves us forward, then halts us, with only the use of her right arm. The queue dutifully obeys these wordless commands.

I reach the point where the orderly queue awkwardly passes the store's exit. At first I do not notice the man leaving the supermarket, but then I realise he's walking backwards. A tall security guard is escorting him out. 'Don't touch me'! the man shouts. 'Don't fucking touch me! Don't fucking touch me! Don't fucking touch me'! he screams. The security guard puts his hands up, as if to nudge the man out of the exit. The man jumps back, and yells some more. The security guard says nothing; he does not touch the man; but, he keeps moving half steps forward, shepherding the man out. Now, another security guard appears. And some supermarket employees. They say little, but make 'calm down' gestures with their hands. The man yells: 'You've got anger issues, you. You've got fucking anger issues. Fucking anger issues, you'! The second security guard intervenes, says something in the ear of the first security guard, who then starts to leave. 'I'll fucking have you', the man yells after him. 'Come and have some of this'! he shouts as he also starts to leave. We, in the queue, who witness this look at each other, over the top of our facemasks: we have not yet learned to communicate with just our eyes, but they all seem to be rolling. Our eyes seem to be saying: 'We live in mad times'. And, I think to myself, the proper response to living in mad times is to be mad. In a short few months, we will learn that security guards are amongst the most vulnerable occupations to COVID-19 – and that black and minority ethnic people are overrepresented in these occupations. It matters that the woman security guard is Black and the men from the store are Turkish and Asian.

Inside the supermarket, there's clear evidence of fear. Vast swathes of the shelves are empty: there's no toilet paper, pasta, tinned foods, surface cleaners of any kind, eggs, flour or paracetamol. People wander slowly past the shelves because this is something to see: it is a sign of the times, so worth looking at. People mutter about 'panic buying', but, of course, the panic buyers are the sensible buyers as they are the ones who anticipated the panic buying. Panic has been normalised. As I leave, a man outside is yelling 'This is Great Britain! Tell the truth! Tell the Truth'! He is holding a black leather-bound book, with gold lettering that I make no effort to read. 'We tell the truth in Great Britain', he screams at no one in particular. I avoid eye contact as he passes. 'Tell the truth'! I hear him shouting as I disappear across the road. As I walk, I listen, but there's no clue to what truth he means. Part of me would like to know, but a larger part is afraid to find out.

I write (and rewrite) in a moment of indeterminacy; we do not know when the COVID-19 crisis will end, nor what it will have done to bodies, affects or politics. People want it over. They want to know when it will end and what the plan is. There is a lot of talk of curves, peaks and plateaus, and second waves; each day, there's an accountancy of the dead, with bar graphs and imagined bell curves. The virus has not told us what its plan is: we cannot reason with it, so it feels like the disaster is the fault of the virus, as if it were a terrorist or a mugger. Yet, it is a mistake to think that the coronavirus is a natural disaster or to anthropomorphise it. That said, we do not yet know what kind of disaster it is. Indeed, it seems to be a disaster many times over. Every death is an individual, a person dying unforgivingly, causing inexpressible loss and grief for untold families, friends, colleagues. So many are dying, so few stories make the news. Yet, we are also told it is an economic disaster. We are told it will change everything. A disaster impacting every corner of our lives. (Although, apparently, it's been good for the planet.

And some are profiting beyond their wildest dreams.) The coronavirus is accreting ever more meanings, as its impacts multiply and intensify. We do not know which of its many meanings will persist and which will not. We do not know how the necropolitics to follow after the coronavirus will play out. We live, right now, not knowing. I am guessing, perhaps hoping, none of the above surprises you.

The coronavirus will teach us many things. Like as not, virology aside, it will mostly teach us what we already know. And I am no different. The coronavirus teaches me that we live in a precarious world, made scary and infuriating by the (extra)ordinary politics of the body and of bodies (from facemasks to lockdowns). But, the world was already fractious and precarious, people already living everyday with crisis after crisis (from floods to droughts, species extinction to financial collapse, from sexual abuse to police brutality), living with deep anxiety and apprehension alongside the propensity for great kindness and generosity. And, I guess, in some small way, this book is a response to the already existing and long-standing 'unsettlingness' of modern life, an ordinary indeterminacy that runs through bodies, through affects and through politics.

This book has been in process for a long time, longer even than the torturous process of writing. There are three things to say about this. First, it is normal to thank specific people when writing academic works. Part of the argument of this book is that it is never that clear where ideas (or feelings) come from. So, I want to thank everyone I have talked to about the matters contained in this book. You have all made some difference to what is here. I admit, in ways that I am probably unaware, and more profoundly I am sure than I know. So, thank you. Second, to contradict myself, I need to thank three people without whom this book would not appear in the world: David Featherstone, whose insights have been incalculable; Jacqueline Scott, whose patience I have sorely tested; and Nadia Bartolini, who has had to endure far too much. Third, I need to acknowledge the source material for certain chapters. Chapter 2 reworks 'Skin, Race and Space: The Clash of Bodily Schemas' in Frantz Fanon's *Black Skins, White Masks* and Nella Larsen's *Passing*, which was published in *Cultural Geographies* in 2011 (pp. 25–41). Chapter 3 draws on 'Spatialities of Skin: The Chafing of Skin, Ego and Second Skins' in T. E. Lawrence's *Seven Pillars of Wisdom*, which was published in *Body & Society*, also in 2011 (pp. 57–81). Chapter 4 recasts 'Beastly Minds: A Topological Twist in the Rethinking of the Human in Nonhuman Geographies Using Two of Freud's Case Studies' by Emmy von N. and the Wolfman, which was published in *Transactions of the Institute of British Geographers* in 2014 (pp. 224–236). Chapter 5 adds a case study of Dora, drawing on a sole authored first draft for the introduction that Paul Kingsbury and I wrote for *Psychoanalytic Geographies* (2016, pp. 8–15) to my chapter in that book, 'A Distributed Unconscious: The Hangover, What Happens in Vegas and Whether It Stays There or Not' (pp. 135–148). Similarly, Chapter 6 removes substantial material from 'Distant Feelings: Telepathy and the Problem of Affect Transfer over Distance', as published in *Transactions of the Institute of British Geographers* (2012, pp. 44–59) so as to add material from the case study of Dora, drawing on a sole authored first draft for the introduction Paul Kingsbury and I wrote for *Psychoanalytic Geographies* (2016, pp. 15–19). I am grateful to Paul for allowing me to use these 'pre-Paul' drafts for this book. Finally, Chapter 8, the conclusion, reworks short passages of material taken from 'The Troubled Spaces of Frantz Fanon' (published in *Thinking Spaces*, edited by Mike Crang and Nigel Thrift, in 2000, pp. 260–277). In general, I have retained the empirical stories within these previously published papers, but they have been re-purposed, up-cycled or re-gifted (depending upon how you look at it). I therefore thank the publishers of the journals and the books for their permission to reprint previously published material.

Every effort has been made to trace the copyright holders and obtain their permission for the use of copyrighted material. The publisher apologises for any errors or omissions in the preceding list and would be grateful to be notified of any corrections that should be incorporated in future reprints or editions of this book.

This book is dedicated to Ben Robinson who, despite his most determined efforts, still suffers from geography.

North London
April 2020

Chapter One
Introduction: Bodies, Affects and Their Politicisation

It is impossible to discuss the relationships between bodies, affects and politics in the abstract: that is, abstracted from the material and ideological conditions of their production, from the processes of politicisation and depoliticisation that bring bodies and affects into, or keep them away from, politics (to paraphrase Harvey, 1993, p. 41). To introduce this book, then, I will start with the story of a particular neighbourhood in West London. It is a story worth telling in its own right, for it involves social murder, as Labour MP John McDonnell put it. However, my purpose is to show how bodies, affects and politics have been entangled at various moments in the area's recent history. But, more than this, I want to argue that there are different regimes of bodies, affects and politics operative in these moments – and it is in the clash between these regimes that different forms of politics can emerge. The problem that animates this book, then, is this: how are we to understand these regimes and what are we to make of them?

Lancaster West Estate, North Kensington

In 1972, work began constructing the Lancaster West Estate in North Kensington, London. The Estate was intended to redevelop part of the Notting Hill area, which had become notorious for its slums, poverty and criminality. This reputation has, for decades, been racialised. Since the HMT *Empire Windrush* first docked (in 1948), the neighbourhood's cheap rooms for rent had proved attractive to new

Bodies, Affects, Politics: The Clash of Bodily Regimes, First Edition. Steve Pile.
© 2021 Royal Geographical Society (with the Institute of British Geographers).
Published 2021 by John Wiley & Sons Ltd.

immigrants from the Caribbean (see Phillips and Phillips 2009). Notting Hill also attracted ruthless slum landlords, such as (most infamously) Peter Rachman. By the 1950s, local white people, especially working-class Teddy Boys, were starting to display hostility towards black people moving into the area. In the summer of 1958, there were increasing attacks on black people as well as the rise of right-wing groups, such as the White Defence League; its slogan, 'Keep Britain White'. On Sunday 24 August 1958, armed with iron bars, table legs, crank handles, knives and an air pistol, a gang of white young men bundled into a battered car and drove around Notting Hill for three hours on what they called – in a ghastly echo of lynching culture in the South of the United States – a 'nigger hunt' ('The Nigger Hunters', *Time Magazine*, 29 September 1958, p. 27). They attacked six Caribbean men in four separate incidents: nine of the gang were arrested the following day in the nearby White City estate, after their car was spotted by police. (Later, in September 1958, to their shock, they were each sentenced to four years in prison by Mr Justice Cyril Salmon.)

The following Friday, 29 August 1958, Majbritt Morrison, a white Swedish woman (who would later author *Jungle West 11* about her experiences) was arguing with her Jamaican husband, Raymond Morrison, outside Latimer Road tube station (which is situated on the western edge of the Lancaster West Estate). A crowd of white people gathered to protect a white woman from a black man (see Dawson 2007, pp. 27–29), despite Majbritt herself not needing nor wanting to be defended. A scuffle broke out amongst the gathering crowd, Raymond and some of Raymond's Caribbean friends. On Saturday 30 August, a gang of white youths spotted Majbritt leaving a dance, recognising her from the evening before they started hurling racist abuse – and milk bottles. Someone hit Majbritt in the back with an iron bar. Yet, she stood her ground and fought back, but, when she refused to leave the scene, the police arrested her. The situation quickly escalated. Soon, a 200-strong mob of young white men was rampaging through the streets of north Notting Hill (half a mile or so to the east of the tube station), armed with knives and sticks, shouting 'down with niggers' and 'we'll murder the bastards' (reported in *The Independent*, 29 August 2008 and *The Guardian*, 24 August 2002, respectively). The mob attacked police with a shower of bottles and bricks. This led to five nights of constant rioting (until 5 September), fuelled by the arrival of thousands of white people from outside the area, and by the retaliation of the local Jamaican population, which eventually armed themselves with machetes, meat cleavers and Molotov cocktails.

Ironically, these events were described at the time as the Notting Hill Colour (or Racial) Riots, implying that these riots were the fault of, and conducted by, black people – when, in fact, black people were the target of white riots. Indeed, it was only the Jamaican fight back that brought the riots to an end, with the police singularly failing to control the situation. Afterwards, the Metropolitan Police refused to acknowledge white racism as a cause of the rioting, despite the testimony of officers on the ground to the contrary. Of the 140 arrested during the riots, 108 were charged with offences, with 9 white youths eventually being sentenced: each was given the 'exemplary' punishment of 5 years prison along with a £500 fine. One response to the riots was the creation of a Caribbean Carnival, first held indoors on 30 January 1959, by Claudia Jones – a Trinidadian activist, who had been deported from the United States in 1955, having famously written about the subordination and struggle of Negro women from a Communist perspective (Jones 1949; see Boyce-Davis 2008). The Caribbean Carnival was an important precursor to the now world-famous Notting Hill street carnival, itself policed as if it were a riot in 1976 and 1977.

Partly as a consequence of the so-called 'Colour Riots', the 1960s saw the north Kensington area embody a reputation for poor housing, drug use, prostitution and violence. Of course, this is characteristically an unfolding story of class and racial inequality, with factions of the white working class remaining antagonistically opposed to the developing Caribbean community, yet with new working- and under- class solidarities being formed across racial lines, through cultures associated with sex, drugs and music. This reputation was consolidated in novels, such as Colin MacInnes' *Absolute Beginners* (1959), which is set against the background of the riots, where race and racism are unavoidable. The area's evident social inequalities and antagonisms also attracted filmmakers.

In 1970, in advance of the imminent destruction of the original street layout by the development of the Lancaster West Estate, John Boorman filmed *Leo the Last* on a set built on Testerton Road. The film dramatically dealt with issues of class and race conflict. In the film, Leo, an exiled prince from a foreign country, becomes a Marxist after he witnesses the exploitation of his poor black neighbours by rich white landlords. Rallying his neighbours together, Leo stages an uprising, quickly overcoming the intellectual classes (in the form of a doctor and lawyer). However, the capitalist class (in the form of rent collectors, shopkeepers and shareholders) proves harder to defeat. Leo retreats to his house. Eventually, Leo is forced to flee, burning down his house (repeatedly) in the process – an uncanny portent of the tragedy to come. Within a couple of years, Testerton Road (along with much of the surrounding area) would be demolished by the wrecking ball of slum clearance and redevelopment. Following Boorman, we might think the wrecking ball represents the inevitable victory of capitalism over the working class, with the antagonisms of race and class flattened by the bulldozer.

Although the Lancaster West Estate redevelopment required the displacement of about 3000 people, few were against the plan to replace the crumbling Victorian housing stock. The original plan was a grand design, involving the creation of a modern housing estate with workplaces, shops, offices and amenities, with improved access to the Latimer Road tube station. The master plan was drawn by Peter Deakins, who had been involved in the first stages of designing the Barbican Centre. Though the grand plan would never be fully realised, building went ahead. The initial phase, starting in 1970, would construct three 'finger blocks' (three- and four-storey housing blocks) and a tower block to the north of the site. The finger blocks had large, enclosed, open spaces with children's play areas.

One finger block, Testerton Walk, replaced the former Testerton Road. The finger blocks were seen as tower blocks laid on their side, with internal walkways to keep the housing as compact as possible and a central covered walkway to provide access. To the north, the finger blocks were anchored by a single tower block, designed by Nigel Whitbread applying principles derived from Le Corbusier and the modernists. Building of the tower block began in 1972 and was completed two years later. The first four of the 24 floor tower contained commercial and administrative units; on the remaining 20 floors, there were 120 one- and two-bedroom units, six dwellings on each floor, to house about 600 people. This building would be named Grenfell Tower (as it stood on Grenfell Road).

London's Burning

By the 2010s, nearing forty years after the Lancaster West Estate had been completed, Notting Hill was best known for its flamboyant Afro-Caribbean carnival, a saccharin romantic comedy film, and massive inequality: popstars, super models and politicians lived in multimillion-pound homes, ordering the latest 'flat white' coffees and quaffing Chenin Blanc wine from South Africa, while the new model estates of the 1970s visibly deteriorated. The area had become trendy, with beautiful and exclusive and increasingly expensive private housing sitting side by side with the rundown Lancaster West Estate. In 2012, Westminster Council began an £8.7M renovation of the Grenfell Tower, which received new windows, a new heating system and, on the outside, aluminium cladding was introduced to improve the block's appearance and rain-proofing. The renovation was completed four years later in May 2016. A new story of class and race had been set in motion.

At 54 minutes past midnight on 14 June 2017, the emergency services received the first reports of a fire at Grenfell Tower. Starting in a faulty fridge-freezer on the fourth floor, the fire quickly engulfed the Tower Block. The fire burned for 60 hours, despite the attendance of 70 fire engines and over 250 firefighters. The fire killed 72 people, in 23 of the tower's flats, mostly above the twentieth floor.

On 21 May 2018, the Grenfell Tower public inquiry began, after completing its procedural hearings in December 2017. (Complete proceedings are available online at grenfelltowerinquiry.org.uk and on YouTube.) It opened with a commemorative hearing, with testimony from the relatives of all the dead. Along with the memories and feelings of the relatives, the inquiry included pictures and videos. There are many stories in the fire – all are heart-breaking.

Marcio Gomes was in tears as he recalled the excitement that news of his wife's pregnancy had brought the family. Hours after the fire, he was holding his stillborn child in his arms, while his wife and two daughters lay in a coma having escaped from their twenty-first floor flat at around 4 in the morning. He told the hearings on its first day: 'I held my son in my arms that evening, hoping it was all a bad dream, wishing, praying for any kind of miracle…that he would just open his eyes, move, make a sound'. The family had plans for Logan; he was going to be a superstar; he was going to be a football fan, supporting Benfica and Liverpool. Marcio added, 'He might not be here physically, but he will always be here in our hearts forever [...] Our sleeping angel he was. We let him go with the doves so that he can fly with the angels. We are proud of him even though he was only with us for seven months'. Later that day, the West End Final Extra edition of the *Evening Standard* chose Marcio's words for its headline: 'I Held My Son In My Arms Hoping It Was A Bad Dream', reinforcing this with a picture of Grenfell Tower in flames. This front page replaced the earlier West End Final headline: 'Grenfell: Don't Make Us Wait 30 Years for Justice', which (curiously) had no accompanying pictures of the tragedy.

The change in the headline might, on the surface, seem innocuous for both are highly emotional: one headline speaks to the families' angry demand for justice, while the other picks up on the families' unbearable loss. These two stories have the same source – yet, in this moment, the unrelenting anger that inhabits the demand for justice is replaced by the unspeakable grief and horror of the tragedy. Perhaps, maybe, because a story about the tragic loss of a child would have more appeal for the *Evening Standard* (a free paper that relies on advertising revenue) than the demand for justice. Yet, although the headlines both draw on Marcio's words, the switch in headlines represents the first signs of the separation of different strands of the Grenfell story: with the raw emotion of unbearable loss becoming detached from the rage-filled demand for justice.

That said, in this moment, anger and grief and hope and love and justice and truth are not yet cauterised from one another. Listen to Emanuela Disaro, mother of Gloria Trevisan, a young Italian architect who was trapped on the twenty-third floor by flames coming up the single stairwell. In a phone call on the night of the fire, Gloria had told her mother, 'I am so sorry I can never hug you again. I had my whole life ahead of me. It's not fair'. Speaking through a translator, Emanuela told the inquiry on 29 May 2018 that she had taught her children not to hate, but that she felt a lot of anger: 'I hope this anger is going to be a positive anger. I would like this anger to help to find out the truth of what happened'. A positive anger would, Emanuela hoped, lead to justice.

Anger, Truth, Justice. Intimately connected. Yet the *Evening Standard*'s West End Final edition headline suggested that this might not be enough: bound up in the demand for justice was the feeling that justice should also mean not having to fight for justice. Maria 'Pily' Burton, wife of Nicholas, was the seventy-second victim of the fire: although rescued by firefighters from the nineteenth floor, she eventually died in January 2018 after months of medical care. Nicholas voiced the concerns of many of the bereaved. He told the *Standard*: 'We should not have to fight so hard to be heard, but every step of the way so far has been a battle. You look at the Hillsborough families, still suffering after almost 30 years. We have to make sure we don't have to wait 30 years for justice' (*Evening Standard*, 21 May 2018 West End Final, p. 4; West End Final Extra, p. 5). The demand for justice and truth may be embedded in and motivated by anger, grief and suffering, but it also reaches out to feelings of sympathy, generosity and urgency. Or, their lack.

Nicholas Burton's reference to the Hillsborough families is highly significant. Hillsborough refers to the deaths of 96 Liverpool fans, crushed to death on the Leppings Lane terraces of Sheffield Wednesday Football Club on 15 April 1989. In fact, if justice is the punishment of those held to be responsible, the Hillsborough families will never see it. Criminal trials associated with the disaster only began on 14 January 2019. Indeed, as a forewarning for those seeking justice for Grenfell Tower, the original intention to prosecute six people, including four senior police officers (including the chief constable), weakened. In the end, only two of the six faced trial: David Duckenfield, the police officer in command on the day, and Graham Mackrell, former Sheffield Wednesday club secretary and safety officer; charges including gross negligence manslaughter (Duckenfield) and breaches of safety regulations (Mackrell). The outcome of the trial, on 3 April 2019, was under a fortnight short of the thirtieth anniversary of the tragedy. While Graham Mackrell was found guilty of a single health and safety charge, and later fined £6,500 (with £5,000 costs), the jury could not agree a verdict for David Duckenfield. At his retrial in November 2019, after over 13 hours of discussion, the jury eventually found Duckenfield not guilty. Afterwards, the Hillsborough families asked a simple question: given that the 96 people killed at Hillsborough were found to have been unlawfully killed, who will be held to account? No one will ever be convicted of their deaths.

In the immediate aftermath of the Grenfell fire, the families and survivors were already acutely aware of the struggle for justice in the wake of the Hillsborough disaster. Indeed, groups from both tragedies would meet each other and share some of the same lawyers. In many ways, theirs is a shared struggle over truths: not just over what happened and whose testimony counts, but also over how these truths are to be interpreted and contextualised.

Justice for Grenfell

After completing the commemorative hearings, on 4 June 2018, the Grenfell inquiry began to hear evidence about the events that led to the tragedy. The fire had begun when a Hotpoint fridge-freezer had caught fire in Flat 16, Floor 4, where Behailu Kebede lived. It was Kebede who had first called the emergency services, asking them to come 'quick, quick, quick'. Representing him was a lawyer who had also represented families at the Hillsborough inquiry, Rajiv Menon QC. Towards the end of a prepared statement, Rajiv Menon rounded upon the remit of the inquiry. He noted that the judge, Sir Martin Mason-Brick, was unlikely to reverse his earlier decision and take the wider context into account. However, Menon was not to be deterred. It is worth hearing him at length (video is available at www.youtube.com/channel/UCMxYjfZsqLa8DanN0r2eNJw. Extract from minute 30:30 to 35:51):

There are certain stark irrefutable facts that one cannot simply ignore about the underlying social, economic and political reality and conditions that culminated in 71 people dying from smoke and fire in a high rise residential building, and a seventy-second person dying a few months later, in one of the richest boroughs in one of the world's great cities in one of the richest countries in the twenty-first century. It is no coincidence that this fire occurred in a building consisting of social housing and former social housing purchased under the right to buy scheme and not in one of the posh swanky high-rise residential buildings around London that cater to the extremely wealthy. It is no coincidence that this fire occurred in a building owned by a Tory flagship borough that has been at the forefront of promoting austerity cuts and deregulation and promoting business and profit over health and safety....

Off camera, someone in the audience shouts 'Justice for Grenfell'; others clap in support. The presiding judge, Sir Martin Moore-Brick, turns to the audience and solemnly insists that there will be no (further) interruptions of proceedings. Menon continues:

... It is no coincidence that the vast majority of the residents of Grenfell Tower were first or second-generation migrants and refugees, the remaining residents being largely local people with long-standing roots in the north Kensington area. Amongst the 72 that died, 23 countries and more were represented. So, race and class are at the heart of the Grenfell story whether we like it or not, whether the inquiry acknowledges it or not, whether the terms of reference are extended or not. Consequently, we say that what happened at the Grenfell Tower in the early hours of June last year was as political as it gets and symbolic of a deep inequality in our society.

The parallels between the Grenfell Tower and Hillsborough tragedies are not connected to the high number of victims, nor to the length of time that public inquiries take, nor to the uncertain possibility that anyone will ultimately face what Menon calls 'real justice' and 'real accountability'. The parallels lie in deep and persistent inequalities in society, especially around class. Indeed, for writers such as Gordon Macleod (2018) and Stuart Hodkinson (2019), the core of this inequality is to be understood in the context of a systematic neoliberal assault on the welfare state, which has rendered public housing marginal, neglected, devalued and stigmatised. They point to the way that the local council, the Royal Borough of Kensington and Chelsea, ignored regular complaints by the residents of Grenfell Tower about its safety, including the lack of a sprinkler system. To this is added the actions of an austerity-driven ruling Conservative Party and unscrupulous private contractors. This view positions Grenfell in a long history of neoliberal assaults on the working class, especially on conditions of work and housing, since Thatcherism in the 1980s (see de Noronha 2019; and also Radical Housing Network, Hudson and Tucker 2019). Indeed, Hillsborough is easily seen as part of this assault.

As importantly, there is a shared struggle not only to voice the injustice, but to have their voices heard and recognised. Those voices are heard in moments.

All too often, justice is undermined by a logic of blame: the claim for justice – formed around a much wider experience of inequality and injury – reduced to a demand for someone to be held responsible. Just as the police had sought to blame the victims, Liverpool fans, for their own deaths at Hillsborough, so the finger of blame has begun to be pointed at the firefighters for their supposed failure to evacuate people from the burning tower.

The parallels in the aftermath of the tragedies also lie in the difficulty of making anger, truth and justice stick together, especially over long periods of time. The parallel lies in the ease with which political institutions can detach heartbreak from anger, anger from the demand for justice, and justice from the political will to change public institutions, such as the police (Hillsborough) or the council (Grenfell): for example, by circumscribing the terms of reference of inquiries. Facts and affects are cauterised from one another: justice de-politicised by its gradual assimilation into the legal process. Thus, broader questions about what justice looks like – which are as political as it gets – are transmuted into narrow socio-technical questions, about cladding, about cost efficiency, about sprinkler systems. These are, of course, important issues, but this transmutation effectively converts the politics of social change into a politics of small changes.

What replaces justice is, as we have seen, heart-breaking. Yet, the over-riding heart-break of the tragedy can itself shape how we understand what went before. It can be easy only to see the tragedies that led up to the tragedy, making the tragedy appear inevitable, the only possible outcome of all the injuries and inequities that went before. Yet, Grenfell Tower, as a microcosm of London life, has more than one story to tell.

Tower of London

For most news outlets, including the BBC, and the Inquiry itself, the single most important story about Grenfell is the story of the fire and its victims: its causes, its shockingly quick spread up and around the tower, the horror of the escape... and tragic, heart-breaking, unbearable death. Yet, out of this story emerges another story: the story of a tower that was teeming with life, giving us a glimpse of a different kind of London – not broken, but getting along. To show this, let's turn to the BBC's remarkable reconstruction of the twenty-first floor (which was chosen by the BBC because it was emblematic of the fine line between life and death) for a *Newsnight* special, put together by Katie Razzall, Sara Moralioglu and Nick Menzies, broadcast on 27 September 2017 (see www.youtube.com/watch?v=-cY_fgeeCzc). The report contains interviews with residents, talking about their beautiful homes. A BBC News webpage recounts the story:

> Two IT workers. A civil engineer. A hospital porter. A charity worker. A management consultant. A supervisor in a clothes shop. A market stall employee and part-time

student. A beauty salon owner. A retired waitress. Ten adults and five children lived on the 21st floor of the Grenfell Tower. Nine of them survived the fire. Six of them – and one unborn baby – perished. (www.bbc.co.uk/news/resources/idt-sh/ Grenfell_21st_floor. Last accessed July 2020)

As Rajiv Menon observed at the inquiry, the story of Grenfell was wrapped up in histories of class, race and social inequality (see also Shildrick 2018). After the fire, these grand narratives were replaced with ordinary stories about family, home, work and living alongside different people. Like all other floors (above the fourth floor), the twenty-first floor was organised into four two-bedroom flats and two one-bedroom flats. The two-bedroom flats each had a corner of the tower, with the one-bedroom flats sandwiched between them. All had access to a central lobby, with lifts and a single stairwell. The Gomes family – Marcio and Andreia and their two primary school-aged daughters, Luana and Megan – lived in Flat 183. Like many residents, they enjoyed living in the tower. The flats had big rooms for the family, but the kids would also play in the tower's communal spaces. Marcio and Andreia are first generation migrants from Portugal. They had lived in the tower for 10 years, but still considered themselves 'newcomers', as many of their friends had been in the tower more than 20 years. Marcio works in IT, while Andreia is a supervisor in a clothes shop. In the tower, Marcio observes:

The diversity was great, you'd meet all sorts of different people – Irish, English, Arabic, Muslim, Portuguese, Spanish, Italians. You'd get to see different cultures. You'd go up in the lift with lots of different people; you'd talk; the kids would play. The tower itself was a community; it was very family oriented. It's been portrayed as a poor tower, a broken tower. It was far from that.

The Gomes family escaped because they were woken by Helen Gebremeskel and her 12-year-old daughter Lulya, who lived in Flat 186, a two-bedroom flat diagonally opposite Flat 183. Helen Gebremeskel was born in Eritrea and arrived in the United Kingdom as an asylum seeker when she was a child. She had been living in the tower for 20 years, but only moved into Flat 183 in 2014. She had spent the last few years renovating and decorating her new home. Helen Gebremeskel is the managing director of, and hairstylist at, H&G International Hair and Beauty Salon. Lulya's best friends were the Choucair family, which lived on the twenty-second floor. On the night of the fire, they had called Bassem Choucair to tell them to get out. The Choucair family did not get out; all died in the fire.

Helen Gebremeskel decided no one was coming to help, so made the decision to leave, against the advice of the emergency services (following accepted procedure). By the time she and Lulya went to the Gomes' flat, the fire had already been burning for over half an hour. Thick black smoke forced them all back into the Gomes' flat. For two hours, they attempted to keep the smoke out: they opened the windows, ran the bath and shower water. At around 3.30 a.m., the Gomes bedroom caught fire. The six of them wrapped themselves in wet tea towels and sheets to make their escape. Although they made it down through the horror of the stairwell, Lulya, Luana, Megan and Andreia were put into induced comas and treated for cyanide poisoning. This is when (as we heard above) Andreia's baby boy, Logan, was stillborn.

In Flat 182 lived the El Wahabi family, which had turned their flat into what neighbours described as a 'mini Morocco'. The father, Abdulaziz, worked as a porter at University College Hospital, while his wife, Faouzia, worked for the Westway Trust, which seeks to utilise the space underneath the Westway flyover for community benefit. They lived with their three children: Yasin, a part-time accountancy student at Greenwich University; Nur Huda, who had just completed her GCSE exams and loved football; and Mehdi, who attended a local primary school and enjoyed judo. Helen Gebremeskel had seen the family escaping at around 1.30 a.m., but then the family had opted to return to their flat. They called the emergency services, which told them to stay in the flat. All five died in the fire. Abdulaziz's sister, Hanan, and her family lived in Flat 66 on the ninth Floor, but they all made it out.

On the twenty-first floor, there were two one-bedroom flats. According to Razzall, Moralioglu and Menzies, the occupancy of Flat 184 was uncertain. They reported rumours about a woman from the Philippines, illegally in Britain, living there. It later emerged that the flat was occupied by Mustafa-Sirag Abdu (see news.channel4.com/2017/grenfell-tower/). He is a civil engineer from Ethiopia. He left the moment he heard about the fire. Ligaya Moore, a 78-year-old grandmother, lived in Flat 181. She had lived alone since the death of her husband, Jim, several years ago. She had moved to London from the Philippines in 1972 and had worked as a nanny and a waitress. No one saw her that night. It only adds to the tragedy that her absence became the stuff of rumour; rumour with a racist tinge.

Flat 185 had been bought from the Council under the right-to-buy scheme in 2000 by the tenant Nigerian-born Tunde Awoderu. A few years later, Awoderu moved out, and the flat became private rented accommodation. He was now vice-chair of the Grenfell Leaseholders' Association. Flat 185 was one of 12 privately owned flats in the tower. Eleven days before the fire, Stewart Lee, a 29-year-old IT engineer, and boyfriend Julian Ng, a 30-year-old management consultant, moved into Flat 185, paying the over £1700 a month rent between them. Both Stewart and Julian were born in Britain. On that night, Julian was working in Crawley. It was Stewart's birthday, so he decided to take the day off and to stay with Julian at his hotel. At 3 a.m., Tunde called them, very concerned about their safety. In shock, they watched the tragedy unfold on their smartphones (as reported in the *Pink News*, 14 June 2017).

These are stories about previously unheard voices, speaking in unfamiliar ways. Tower blocks are overwhelmed stories of poor housing, unemployment and social deprivation. Indeed, as Ida Danewid has shown, placed in an international context, Grenfell can be seen as a consequence of wider forces of racial capitalism that is reconfiguring colonial and postcolonial relationships within London (2019). From this perspective, Grenfell is more than just 'Empire coming home', it is about the way racial capitalism is shifting people and capital around the world, involving a racist global necropolitics in which the lives of black people do not matter. Indeed, the relationships within the tower were diverse, geographically; they stretched to places far away, to Morocco and to Ethiopia, to Italy and to the Philippines. Yet, Grenfell also domesticated the global: showing how migrants had settled in London, revealing migration to be homely and familiar, ordinary (www.bbc.co.uk/news/world-44395591). The testimonies of those living on the twenty-third floor, whether they intended to or not, made it clear that these homes were beautiful, the people in the blocks were hard-working, and they had created bonds of friendship and community. Evidently, too, this was not true for everyone: some people were isolated, not able to turn to friends, their families far away. Grenfell holds more than one story, at the same time.

People had settled into London life, working nearby, going to school, moving within the tower and within the neighbourhood. The tower block was diverse, socially: alongside young urban professionals were public sector employees and charity workers. A system of 'self-evident facts' (as Rancière puts it, 2004, p. 7) about tower blocks was being systematically shaken: the block's many jarring stories were demanding to be heard. In Rancière's terms, this amounted to a challenge to the distribution of the sensible: that is, a regime of perception and understanding through which people, ideas and things are assigned a 'proper place' in relation to other people, ideas and things. In particular, dominant narratives of class and race – intimately connected to notions of social housing, refugees, migrants and tower blocks – were being directly challenged. Not just through stories about community and hard work, but affectually through stories of love, family, hope and unassimilable trauma and grief (see also Doherty 2018).

Following Rancière (2000), we can also argue that, in the aftermath of the fire, bodies were torn from their assigned places and a form of free speech and expression emerged. For a few days, these bodies caused a disruption in the ordering of relations of power, not by revolution or riot, but by affect: that is, affect was having political effects (following Massumi 2002, p. 40). The hurt and pain and loss and anger simply overwhelmed the political response, with its focus on emergency services, its desire to convert the fire into a set of technical questions of fire-proof properties of cladding and architectural design, and into amounts of money (£5 million, up front). These affects cut across forms of cultural and religious identity, altering questions of belonging and hierarchy: the Grenfell Tower fire brought new subjects into the field of perception. Working class, migrant and refugee, and community voices were being expressed, heard and spread. Grenfell, then, is a story about voices lost – and voices gained. More, perhaps most importantly, the story of the tower is also a struggle over what justice looks like.

As Political as It Gets

Grenfell's stories of loss, pain and grief are exacerbated by another story: the failure of the council and government to respond adequately to the tragedy, both in the days after and the following months, with survivors still living in temporary, unsuitable accommodation. In this story of Grenfell, venal political indifference is pitted against the strength of the community (see Madden 2017).

As people were being taken to hospitals across London, as the fire was still burning, and as the council failed to comprehend or respond to the scale of the tragedy, people from the surrounding area were already starting to donate food, water, bottles, blankets, clothing, nappies, toys at St Clement's Church, the Al Manaar Muslim cultural heritage centre, the Notting Hill Methodist Church and the Westway Sports Centre as well as Kensington Town Hall. As the council floundered in its response, the communal response was immediate…and it was as compassionate as it was angry. These were the resources out of which a strong community emerged. Politicians were keen, too, to show how much they cared.

Leaders of both the major political parties visited the Grenfell Tower scene on the morning of 15 June. At 10 a.m., Conservative Prime Minister Theresa May made what was called a 'private visit', meeting members of the emergency services. A couple of hours later, Jeremy Corbyn, the leader of the Labour opposition, arrived at the scene. The contrast between the two visits could not have been starker, and people were quick to point this out on social media. On BBC TV news, images of Jeremy Corbyn listening to local people voice their anger and putting his arm around a distressed woman inside St Clement's Church were juxtaposed with Theresa May's seemingly dispassionate conversations with groups of firefighters and police. People were quick to criticise Theresa May for not meeting residents, for not being prepared to take their anger and pain. More, she was condemned for talking to people via TV about how the government would do everything it could to help and rehouse them. She was being condemned as much for not listening as for not talking; as much for failing to be open to people's anger and pain as for a seemingly dispassionate display of compassion. Perhaps in an attempt to seem more human and humane, just before 2 p.m., Theresa May announced that there would be a full public inquiry. During the morning, the official death toll figures had risen from 12 to 17. No one thought the death toll would stay at 17; many thought it would rise into the hundreds. The sense that officials were underestimating the tragedy only intensified the anger on the streets around Grenfell Tower.

While Theresa May was announcing the public inquiry, freelance journalist Mario Cacciottolo, writing for the BBC, was reporting the views of local residents (http://www.bbc.co.uk/news/uk-40291372). He spoke to Maria Vigo, who had lived opposite the tower for 11 years. She told him: 'There was a lot of anger on the school run this morning. There's a lot of separation between the classes, and people are telling me that it's down to social cleansing'. She complained about being priced out of the area by the rising cost of basic amenities. She blamed gentrification. 'This area's always been working class. It's starting to become a bit less so now, and the working class are feeling that they're being left without a voice. The council isn't listening to us. We don't want a pretty building. They should ask us 'what do we need'? Or 'what would we like'? In the rich borough of Kensington and Chelsea, Grenfell Tower stood in an area (surrounding the Latimer Road tube station) that ranks amongst the poorest 10% in England.

As Mario Cacciottolo wandered around the streets surrounding Grenfell Tower, he discovered another story. He spoke with Christina Simmons, who had lived near the tower for 27 years. She explained: 'People are coming together and rallying together. I didn't realise we had so many Eritreans and Somalians; they've come out to offer support'. The growing pride in the community response did not mask the problem. Christina added, 'They don't listen to us. We're being neglected and ignored. I'm bloody angry'. Then, on a more hopeful note, 'Maybe this'll bring us all together'.

A day later, the relief effort of the council had not improved. The communal effort, meanwhile, was now enormous, with makeshift relief centres at the Westway Sports Centre and St Clement's Church straining under the weight of goodwill. While the Queen and Prince William calmly met volunteers, local residents and community representatives at the Westway Sports Centre, a large angry crowd gathered outside St Clement's Church, where Theresa May was due to meet victims of the disaster. As she appeared, clear on the TV reports were shouts of 'coward' and 'shame on you'. Some broke through the police cordon to shout at the Prime Minister as her car whisked her away. It is not clear whether Theresa May heard them.

The fire at Grenfell Tower exposed the growing social inequality in the area and tensions around gentrification as well as the indifference and unpreparedness of the local council and the Conservative Government (although they would protest otherwise). Yet, the fire also changed the way the area was seen and heard. Perhaps the most remarkable shift was prompted by the communal response: the tragedy operationalised a community that was previously latent and fragmented. Indeed, the escalating acts of compassion crystallised what appeared from the outside to be an impossible community. Significantly, community was quietly articulated through faith-based organisations – especially St Clement's Church and the Al Manaar Muslim heritage centre – and expressed (informally) in religious terms (see, e.g., Everett 2018).

On a wall of the Latymer Community Church, in an echo of other tragedies in the United Kingdom and elsewhere, people wrote tributes and messages, and left flowers (see www.getwestlondon.co.uk/news/west-london-news/gallery/tribute-wall-grenfell-tower-community-13191587; and, www.standard.co.uk/news/london/hundreds-of-harrowing-messages-left-on-wall-at-grenfell-tower-fire-scene-a3565521.html). At the centre of one panel, underneath a wicker heart, in multicoloured felt pen, were the words: 'Pray for our Community'. Surrounding these words were messages of love, grief and hope, written in English and Arabic. Many of the messages were drawn upon religious themes, practices and iconography to articulate a sense of community: 'No words can truly describe the pain, the fear, the sense of being lost – only emotions to be felt – the heart felt prayers of the whole world are here and with time and peace will heal and reunite loved ones'. Another read: 'What a wonderful response to a horrible tragedy. May we continue to give so that no one is in lack. This is the kingdom of Jesus Christ'. Alongside it, someone had written: 'May Allah grant all the highest ranks in jannah, and all the survivors the strength to carry on'. It wasn't just that the fire had revealed the different faiths in the community, it was that the commonalities amongst the seemingly incommensurable faiths had become the grounds upon which people were acting together. Theologically different ideas about the soul and heaven found common ground in the articulation of a desire that people's souls would find heaven and rest in peace. Gifts and giving were at the heart of a now operative impossible community, underpinned by expressions of love, hope, help, resilience, sympathy and strength.

The operability of community was often expressed through ideas of togetherness and unity: 'Bonds formed in fire are difficult to break – our community will always stand together'. And, on another panel, simply: 'WEARE1'. On the other hand, community was also what set people apart from the rich and the powerful: 'May their souls rest in peace #PPPpoorpeoplepolitics', 'People help the people'. And, there was also a palpable sense of anger and a desire to see justice: 'Why did this need to happen?', 'Theresa May where are you, your people voted you in, Justice for Grenfell, Jail Those Responsible'.

The operative community quickly was becoming two-sided: one side, joining the people together, while at the same time separating the people from the rich and the powerful. The demand for justice similarly faces in two opposite directions at once. One demand is faced towards the law. This is the demand for the causes of the tragedy to be identified and the people responsible to be held accountable: this is what Rajiv Menon calls 'real justice' and 'real accountability' (see above). Theresa May's quick declaration that there would be a public inquiry addresses this demand: yet, the public inquiry also utilises the limits contained in this demand for justice – the inquiry's remit would not extend to wider questions of race, class and social inequality. Yet, that circumscription of the inquiry itself creates a contestable limit. Another demand faces away from the law as the law itself is the source of injustice. This demand for justice chains the tragedy to longer histories of racism, social inequality and deprivation. It is glued to a sense of injustice that calls into question the meaning of justice. In this, Hillsborough is a marker of what justice does not look like. It does not look like the legal process and the rule of law. It does not mean years of struggle against 'the system' by heart-broken, nearly exhausted, resource poor, working families. It does mean taking into account – and redressing – long histories of class, race and wider social inequalities (as Gordon Macleod, Stuart Hodkinson and Ida Danewid, amongst many others, argue).

Tell Me What Community Looks Like

In his analysis of a fire at the Imperial Foods plant on 3 September 1991 in Hamlet, North Carolina, David Harvey observes:

> We here encounter a situation with respect to discourses about social justice which closely matches the political paralysis exhibited in the failure to respond to the North Carolina fire. Politics and discourses both seem to have become so mutually fragmented that response is inhibited. The upshot appears to be a double injustice: not only do men and women, whites and African-Americans die in a preventable event, but we are simultaneously deprived of any normative principles of justice whatsoever by which to condemn or indict the responsible parties. (1993, p. 53)

The parallel between the Imperial Foods fire and the Grenfell Tower fire is uncanny. Yet, it is also misleading. In fact, the inquiry – which is capable of identifying those responsible (as with the Hillsborough inquiry) and making recommendations for changes (such as, most likely, new fire regulations regarding cladding materials) – focuses the political response on technical matters: thus, 'real justice' and 'real accountability' are exactly what can be delivered, while looser, more unruly, notions of justice become marginalised and silenced: literally, as Moore-Brick's rebuke of the audience at the inquiry demonstrates (see above).

Brilliantly, the community has turned these silencing strategies on their head. Silence has become a means through which to remember the tragedy, to reconfirm a commitment to remember what happened and to struggle for justice (see Charles 2019). On the fourteenth of every month, since the fire, there has been a 'silent walk' to pay respects to those who died. It starts at Notting Hill Methodist Church and follows a path towards the Lancaster West Estate. On the first anniversary of the fire, not only was there a nationwide minute's silence in the morning, there was also a silent procession at noon to the 'wall of truth' at the Maxilla Social Club, followed by an afternoon silent procession to the Methodist Church, ending with an evening silent walk starting at the wall of truth. The silent walks are more than an act of remembrance, however. They express the solidarity of the community; they also symbolise the way that community voices continue to be ignored (see www.grenfellconnect.org.uk). Significantly, after the walks, there were also prayers and remembrance at the Al Manaar Muslim Cultural Heritage Centre.

In contrast, the Justice4Grenfell campaign, along with the Fire Brigade's Union, organised a more conventional protest march on the Saturday, 16 June 2018, immediately following the silent walk that marked the first anniversary of the tragedy. Assembling outside Downing Street (residence of the British Prime Minister), the march began with speeches decrying the council, the government and the contractors who installed the cladding on the tower. More importantly, the speakers each affirmed the need to look back over the longer history of the tower. They talked about how the tower had been built to high standards, yet had been left to decline through decades of under-investment under successive political regimes. They vented fury at austerity urbanism that the Tory-led council had embraced with enthusiasm – focusing its attention on profits, rather than living standards and safety. The silence was powerful, the speakers affirmed, but they wanted this protest to be noisy. From Downing Street, the march wound its way past the Houses of Parliament. The Houses of Parliament are currently undergoing a £3.5~7Bn refurbishment, ironically, partly prompted by fears that the risk of fire is so great that the building is a death trap (reported by *The Guardian*, 31 January 2018).

At the Home Office, the government department responsible for overseeing the conduct of councils, the march paused. Here, speakers condemned the government's politics of race, which casts migrants as a problem and as unwelcome in the United Kingdom. Speakers pointed to the Muslim community that had been the first to respond to the fire, running towards the fire while the politicians ran away from it. The politics of justice, in this form, pits itself not just against the injustices perpetrated by Conservative Governments, but against the politics of government in general, Labour and Conservative, that have failed 'the community'. As the marchers left, they chanted: 'Tell me what community looks like/This is what community looks like' (Figure 1.1).

Figure 1.1   The Justice for Grenfell March to mark the first anniversary of the Grenfell Tower fire, 16 June 2018, as it passed the Houses of Parliament a second time (on the way back from the Home Office). Source: Steve Pile.

The speakers on the march drew clear parallels between the anti-migrant discourse of Brexit (on both the remain and leave sides), the scandal of the Windrush generation (who had become collateral damage in Theresa May's 2012 creation of 'a really hostile environment for illegal immigrants' in the United Kingdom, but especially from early 2018 onwards) and the systemic underinvestment and marginalisation of the Grenfell community (see Bradley 2019; and, El-Enany 2019). So, when the march asked what community looked like, it was not simply making visible an affective community forged by compassion, grief and anger, it was pointing out that the body of the community was forged out of racial and religious diversity. This community incorporates invisible and visible differences.

The Grenfell Tower tragedy altered a pervasive narrative about Muslims and community in London that had become dominated by suspicion and fear. For example, in the wake of multiple terrorist attacks – an attack at London Bridge by three Muslim men less than two weeks before the Grenfell Tower fire had killed eight people – Muslim communities were constantly being asked to account for their relationship to Islam. The Grenfell Tower tragedy uncouples the relationship between Islam and terror in two significant ways. On the one hand, the stories of the survivors are about the lives of ordinary Londoners: making their homes beautiful, working hard, getting on, looking after family, and living in a community of friends and strangers. This ordinariness undoes a divide between Muslims and everyone else. This is evident in the 'wall of truth', which does not discriminate between faiths and backgrounds (see www.bbc.co.uk/news/ resources/idt-sh/grenfell_tower_wall). On the other hand, Muslims were prominent in organising and providing support. The Al Manaar Muslim Cultural Heritage Centre is now a symbol for Islamic generosity, aid, support, community. Consequently, when Muslim Aid produced a report in May 2018 praising the voluntary sector for its response to the fire, while also being critical of the council response, it was widely reported by national news media (see, e.g., www.bbc.co. uk/news/uk-england-london-44295397).

In the wake of the Grenfell fire, we can see that the normal conventions through which local people are rendered visible and invisible, heard and unheard – and, the common frames through which their bodies and lives are understood and interpreted – were, however temporarily, suspended. Instead, unheard stories were told and listened to: stories about migration, race, class, family that all challenged received understandings about tower block life and the composition of community. In part, the shock to the system turns around the event itself and the affective intensity of the moment. Yet, we can also see the ways that affects enabled the crystallisation of an operable community, forged around a shared sense of injustice and a demand for justice. It is therefore worth noting that one of the first marches in London, on Sunday 31 May 2020, to protest the death of George Floyd (in Minneapolis) ended at Grenfell Tower. In silence. As someone wrote Black Lives Matter on the memorial wall.

Body Politic

Before the fire, there were long histories of class, race and social inequality that were all refracted through notions of violence and poverty. Grenfell exposes these: its anger is intimately connected to them. Yet, it also exposed the ordinary affects of family, work, humour, hope, childhood and so on (see Highmore 2011; and Stewart 2007). Communal solidarity formed around the tragedy, not only through anger and feelings of sorrow and generosity, but also around these ordinary affects. This underscores Rancière's injunction to see politics as a form of experience, not in its exceptionality, but in its everydayness. This asks us to pay particular attention to ordinary life in Grenfell Tower, as a political moment that invites us to consider how people are assigned to places and to a place in life.

Following Rancière, we can see the emergent stories and voices from Grenfell Tower as a challenge to the distribution of the sensible (2004, p. 8; see Dikeç 2015): that is, a recasting of common sense understandings of race, class and religion. In Rancière's terms, this recasting was forced by marginalised and excluded bodies challenging accepted ways of understanding those bodies, as if from the outside of a bodily regime (to use Fanon's expression) that seeks to define bodies and assign them a proper place in the scheme of things. In these terms, a bodily regime defines, and is defined by, the senses: not just the visible and the invisible, but all the ways that things are sensed or not, understood or not. Bodily regimes are about how the senses are defined and lived. Significantly, bodily regimes are produced and reproduced by the drawing of an epistemic boundary that differentiates between the sensible and the insensible. This boundary enables a distinction between people who are inside a community of shared sensibilities and those who are outside:

> The distribution of the sensible reveals who can have a share in what is common to the community based on what they do and on the time and space this activity is performed. Having a particular 'occupation' [both an activity and a space] thereby determines the ability or inability to take charge of what is common to the community; it defines what is visible or not in a common space, endowed with a common language, etc. (2004, p. 8)

The distribution of the sensible, for Rancière, is as political as it gets:

> It is a delimitation of spaces and times, of the visible and the invisible, of speech and noise, that simultaneously determines the place and the stakes of politics as a form of experience. Politics revolves around what is seen and what can be said about it, around who has the ability to see and the talent to speak, around properties of spaces and possibilities of time. (2004, p. 8)

Let us not romanticise the Grenfell Tower tragedy, nor mark it as some upheaval in the constitution of the political. Yet, the fire, however temporarily, not only illuminated the boundary between who could speak and who could be heard, who was visible and who was invisible, what community looked and sounded like, it also meant that this could be contested…and contested anew. Indeed, a flurry of documentaries on Grenfell made visible and gave voice to the community, the histories of injustice and inequality, and also gave voice to the survivors and relatives (see Sborgi 2019). As Eddie Daffarn, who lived on the sixteenth floor of the Tower, in the 2018 documentary *Grenfell* (by Minnow Films, shown on BBC1 on 11 June 2018), put it:

> I was very proud to live there, and it's just a tragedy that we're, you know, not able to show people, you know, what we were actually like […] Justice. Reparation. These words are really meaningless without fundamental change. Grenfell can represent a turning point in history. A change in how society views communities that live in social housing. That, they are listened to and respected, because had we been listened to and respected at Grenfell, it would not have happened. It's as simple as that.

These kinds of justice, rooted in life itself, are justice from outside a political regime that defines justice from above, that constricts justice in and through the legal system and public inquiries that focus on socio-technical systems. But they are also coexistent with it.

In this light, disturbances in the distribution of the sensible may not be the consequence of those outside the regime suddenly becoming visible to those inside the regime, but a clash between multiple, already-existing distributions of the sensible and regimes of the body (see also Rancière 2004, pp. 46–47). This is the analysis that I will pursue in this book. This book explores both the coexistence of bodily regimes, and also the relationship between them. Of course, coexisting bodily regimes (around class, race, gender, sexuality and so on) can be mutually supportive, integrated and coherent. Ideas such as intersectionality have been developed to account for this. However, I am more interested in the ways that bodily regimes fail to 'add up' to a single coherent integrated system of privilege and power. Indeed, I am most interested in those moments and places where bodily regimes clash. This is what I see in the Grenfell Tower tragedy. It exposed the distribution of different voices and bodies, yet it also offers new stories – stories that do not add up to a singular understanding of the event, nor that can be closed down around pre-determined understandings of those bodies, of affects or of politics. Rather, the tragedy reveals the *indeterminacy* of bodies, affects and politics – and it is in this that its radical possibilities lie. Like Hillsborough, Grenfell simply refuses to settle down, to be quieted.

For Rancière, political subjects are produced by fissures in the distribution of the sensible. It is through these fissures that a political process can be unleashed that opposes or resists the sensible order. The sensible order is an organisation – in Rancière's terms, a police order – of bodies into 'a system of coordinates that define modes of being, doing making and communicating that establishes the borders between the visible and the invisible, the audible and the inaudible, the sayable and the unsayable' (Rockhill 2004, p. 93, in Rancière 2004). The sensible order assembles and parcels up people, assigning them to a proper place within modes of perception, but also in modes of knowledge, agency and passivity. The sensible order produces self-evident truths about the way things are: creating modalities of what is visible and what is audible, creating modalities of what can be thought, said, made and done. However, Rancière acknowledges that contradictions can be generated between orders of the sensible, for example, between the visible and the sayable. He terms this silent speech: a hidden layer of meaning, awaiting expression, beneath the orders of speech and visibility.

Rancière's central insight – about the organisation of bodies and the senses into regimes of the visible, the sayable and the actionable – must be altered in the light of Grenfell. As well as seeing hidden layers of meaning, we must rather register those moments when regimes of the body and the sensible clash or rub up against one another. These frictions and clashes occur between regimes that are organised at many different points in the experience of power relations, of affects and of bodies. They are not somehow outside or beyond systems of organisation of bodies, affects and power relations. Rather, these systems are open, porous, incomplete and mutable in ways that consistently fail to close down experience around a singular distribution of the sensible. Here, I follow work by Mustafa Dikeç (2005, 2015) and Divya Tolia-Kelly (2019). For both Dikeç and Tolia-Kelly, Rancière affords the opportunity to include excluded people in an understanding of, respectively, urban politics and the politics of the museum. Significantly, in Dikeç's and Tolia-Kelly's work, those excluded people are further marginalised on grounds of race through the production of space. Thus, for Tolia-Kelly, the exclusion of racialised artists is achieved through a particular colonial distribution of the senses (see also Tolia-Kelly 2016). Dikeç, meanwhile, shows how the built environment is used to marginalise racialised groups, especially immigrants, to an (im)proper place: for example, by housing them on the peripheries of cities or making access to the city's social and physical infrastructure difficult or even impossible. Bluntly, racialised subjects are pushed to the edges of cities and museum spaces – and thereby rendered voiceless, unseen and unintelligible.

Critically, this form of exclusion does not solely operate by policing space such that people are simply placed beyond the city or the museum, rather there is a topological ordering space. This topological ordering of space ensures that people can be socially excluded even while they are spatially proximate (on this, see Allen 2003; and especially 2016). Thus, even while racialised subjects are inside the city and the museum, they are kept socially distant, either by the racial segregation of urban space or through the production of white museum spaces (see also Bressey 2009, 2014). For me, taken together, what Dikeç and Tolia-Kelly reveal is that two orders of the sensible can coexist, even as they are held apart. Indeed, it is their coexistence that ensures that the distribution of the sensible has to be policed, sometimes brutally (see Dikeç 2018) and, arguably, sometimes with housing policy and/or with malicious neglect, as the Grenfell and Hillsborough tragedies show.

More than this, regimes of the sensible are not just organised vertically, from above, but also horizontally, between and amongst people. Significantly, what this brings into view is that the clash of regimes is not just conducted at the border of the different means of organising and policing of who can and cannot speak, be seen or not, can action or not, but is also brought about by the existing together of different ontologies, epistemologies and experiences of speech, vision, action and so on. To show that this becomes operative through bodies, affects and politics is the work of this book.

From this perspective, it is in these clashes within and between regimes of bodies that political subjects emerge. Thus, we cannot see political subjectivities as always already forged and fixed. Rather, regimes of the body, themselves, reveal that bodies remain stubbornly indeterminate, even while they furiously seek to fix and stabilise them into their proper place (as Rancière shows). The enemy of politics, in this view, would be to convert the indeterminacy of the subject into a politics of identity, where all things are already known and fixed in place. Instead, I will seek an account of politics that is grounded in both the overdetermination and the indeterminacy of affects, bodies and identities. By overdetermination, I mean the ways that bodies and identities can be *determined many times over* by structures of meaning and power – which I will refer to, in aggregate, as bodily regimes. Politics can, and does, emerge in opposition to these overdeterminations. However, in this book, I wish to emphasise the ways that politics emergences both through the tensions between various forms of determination and also through indeterminacy.

Understanding Bodily Regimes: Between Rancière and Freud, between Overdetermination and Indeterminacy

As we have seen, the approach I take to thinking about the overdetermination and indeterminacy of bodies, affects and identities draws on Rancière. Rancière's notion of the distribution of the sensible invites an analysis of bodies, affects and politics that focuses on the unconscious ways (which he calls an aesthetic regime) that the bodily senses are structured, such that only certain people are noticed, listened to and understood. I have previously suggested that aesthetic regimes (the unconscious structuring of the sensible) are multiple, inconsistent, mutable and (can) occupy the same space.

The idea that the processes of unconscious structuring are many has three implications that are significant for my approach. First, an account of the unconscious, and unconscious processes, cannot be restricted to singular functions and outcomes, such as repression and the Oedipus Complex (see Chapters 4 and 5). Second, unconscious structurings (plural) of the sensible might be in tension or conflict with one another (see Chapters 2 and 3), albeit in ways that are not easily perceived or that might be opaque or hidden or even repressed. Third, conflicts and tensions within and between unconscious structurings requires a dynamic understanding of unconscious processes, capable of illuminating and understanding the mutability of distributions of the sensible – and, for my purposes, of bodily regimes (as in Chapters 6 and 7). My approach builds on Freud's account of the unconscious, yet this requires some recasting. Thus, Rancière's discussion of Freud is instructive and illuminating – as it, helpfully for me, sets up the analytic architecture of this book.

Rancière's intention, in writing *The Aesthetic Unconscious* (2001), is to understand Freud's use of the Oedipus myth. This discussion does several interesting things for me: first, in answering the question of why Freud chooses the version of the myth that he does, Rancière provides both a demonstration of the effects, and affects, of specific aesthetic regimes and also (I argue) an example of the coexistence of aesthetic regimes; second, this then enables a re-evaluation of the place of the myth in Freudian thought, which allows me to de-privilege Oedipus in my version of Freud; and, thus, third, this opens up new avenues of thought for thinking with Freud about bodies, the unconscious and the distribution of the sensible.

In *The Aesthetic Unconscious*, Rancière boldly asserts that Freud's understanding of the unconscious is predicated upon a particular aesthetic regime (p. 7). This aesthetic regime, for Rancière, creates a very particular set of dichotomies between thought and non-thought, between knowing and not-knowing, between seeing and blindness, between listening and hearing, between logic and sense. To substantiate his argument, Rancière turns to Freud's discussion of the Oedipus myth, which provides Freud with a cornerstone for understanding the sexual anxieties of childhood, especially concerning castration (see Pile 1996, ch. 4). Rancière is especially interested in which version of the Oedipus myth Freud selects. I am persuaded by Rancière's argument that this selection is significant and telling.

Freud's account of the Oedipus myth is first spelt out in detail in the *Interpretation of Dreams*, under the section dealing with dreams about the death of loved ones (1900, pp. 275–278). He briefly outlines the myth, which involves both the unwitting fulfilment of a prophecy and also the vain efforts of people to avoid threatened disaster. Briefly, in Freud's telling of the myth, King Laius of Thebes gives his infant child to a servant to get rid of because of a prophecy that Laius will be killed by his own son. Unable to kill the child, the servant gives the baby to a shepherd, Polybus. The child, Oedipus, grows up ignorant of his origins. Oedipus, later, kills Laius at a crossroad, not knowing that he has killed his father. Then, he marries Queen Jocasta, Laius' widow and also his mother, and becomes king of Thebes. Oedipus' life begins to unravel when a plague hits Thebes. Oedipus sends his brother-in-law, Creon, to the oracle at Delphi to discover the reason for the plague. Creon tells Oedipus that the problem is that the murderer of Laius has never been caught. Oedipus vows to find the murderer. He sends for the blind prophet, Tiresias. Tiresias is blunt: 'You yourself are the criminal you seek', he tells Oedipus. Oedipus simply cannot see how this can be true. In the ensuing argument, Oedipus mocks Tiresias' blindness, and Tiresias retorts that it is Oedipus who is blind. As Tiresias leaves, he mutters that the murderer is son and husband to his own mother, brother and father to his own children, and a native of Thebes. Long story short, through a series of misdirections and revelations, eventually Oedipus comes to know that he is the murderer that he is seeking – and once he comes to a full realisation of his situation, he (famously) tears out his own eyes.

Why this version of the myth, Rancière asks? Other versions were available to Freud: for example, in 1659, Corneille wrote his own version; and, in 1717, Voltaire completed a version while in prison for 11 months (*The Aesthetic Unconscious*, ch. 2). For both Corneille and Voltaire, Rancière asserts, the original myth was not only too incredible, but also too gory for the sensibilities of their audiences. It was just implausible that Oedipus would not know that he was the murderer after being told so, bluntly and unambiguously, by Tiresias. More than this, the tearing out of his own eyes was plainly too literal and too explicit. Corneille and Voltaire sought to create more mystery around the identity of the murderer by introducing new, additional suspects. And they made sure that Oedipus' denouement occurred off-stage, unseen by the audience. Freud selected neither of these versions of the myth, nor indeed other versions; such as Dryden and Lee's *Oedipus: A Tragedy* (1679), which centres on the love affair between Oedipus and Jocasta and portrays Oedipus as noble and heroic. Against expectation, Freud's choice, Rancière argues, has nothing to do with incest (although Voltaire reduces the significance of incest in the story); it is instead about Freud selecting between aesthetic regimes.

In fact, Rancière argues, it would have been easier for Freud to have chosen either Corneille's or Voltaire's Oedipus, for these operate within the dominant representational regime of aesthetics, where characters and situations are taken literally. For both of them, it was simply impossible for audiences to suspend their disbelief when Oedipus cannot believe what Tiresias tells him. They believed their audiences simply would not understand that Oedipus was 'in denial' (as it is now popularly termed). Interestingly, this was apparently not the case for restoration period English audiences, as Dryden and Lee's version followed Sophocles' plot closely. Either way, for Rancière, neither incest nor Oedipus' ability to hear or see the truth is at the heart of Freud's choice. Rather, Freud chooses Sophocles' Oedipus because the classical aesthetic regime (as Rancière calls it) places defects in the subject at the heart of tragedy. This sense of flaws in people's characters provides Freud, in this view, with the ability to imagine unconscious processes, such as denial and castration anxiety.

Building on this observation, Rancière is now able to contrast the classical aesthetic regime with today's dominant regime, the representational aesthetic regime. The pivotal difference between the representational and classical aesthetic regimes is the way they render the truth visible and invisible. Rancière argues that Freud selects the classical regime precisely because Oedipus cannot see his own truth, even when it is pointed out to him. This performs a particular cut through seeing and not seeing, the sayable and the unsayable, hearing and understanding, through knowing and not knowing. On one side, there is the knowing subject, Laius, who seems to be king of his own destiny; and, on the other side, there is a not-knowing subject, Oedipus, who cannot even be king of who he is. It is the movement between these contradictory positions that animates the tragedy and what draws Freud to it. This is significant for Rancière.

And, for me, too. Rather than Oedipus being a lesson in what happens if you break the incest taboo (knowingly or not), this is about the relationship between knowing and not knowing, seeing and not seeing, hearing and not hearing (see Chapters 2 and 3): that is, the relationship between conscious and unconscious thought processes (see Chapters 3 and 4). And, as we see (in Chapters 5 and 6), between a repressive unconscious and a communicative unconscious. In particular, the (classical or representational) aesthetic regime is connected to the ideas of over-determination and indeterminacy, where overdetermination is about how meaning is determined many times over through the distribution of the senses (that creates ways of knowing, seeing, hearing, feeling and so on) and where indeterminacy is a product of the epistemic cut between knowing and not knowing, seeing and not seeing, hearing and not hearing, feeling and not feeling and so on. My approach to these issues is empirical. The relationship between bodies, affects and politics is not to be decided in the abstract or in advance, but in context.

Rancière also asks why Freud draws upon Oedipus at all. That is, why does Freud draw upon a fictitious character from a stage play to model psychic struc-tures? This question can be usefully extended: why does Freud draw on Shakespeare's Hamlet (immediately following his discussion of Oedipus in *The Interpretation of Dreams*) or other artistic products, such as Michelangelo's statue of Moses (Freud 1913), Leonardo da Vinci's painting of *The Virgin and Child with Saint Anne* (Freud 1899), 'The Sandman' horror story by E.T.A. Hoffman (Freud 1919, which also establishes a connection between eyes and the fear of castration), and the like? For me, these examples perform two significant functions. First, these examples, for Freud, bear witness to unconscious processes through the aesthetic forms they take. Second, they reveal that unconscious processes bleed through life, in all its forms: that is, given the focus of this book, through bodies, through affects and through politics. Thus, as this book is to bear witness to the 'aesthetic unconscious' of bodies, affects and politics, I draw on a range of exam-ples: I have selected Freud's case studies (Chapters 4, 5 and 6), autobiography (Chapters 2 and 3), novels (Chapter 2), Hollywood movies (Chapter 5) and art (Chapter 7). Importantly, for my argument, the unconscious is not confined to a particular space, such as the consulting room, and especially not to the brain or the mind of the individual.

The Oedipus myth is normally taken as significant because it places repression, sexual anxieties associated with parents and the body, at the heart of the analysis of the symptom. However, what Rancière highlights is the way Freud traces the slippages between knowing and not knowing, between hearing and not hearing, between seeing and not seeing. So, the imperative that this book follows, drawing on both Rancière and Freud, is two-fold. First, to map out the coexistence of different distributions of the sensible and their policing – which I will call bodily regimes (following Frantz Fanon, see Chapter 2). Second, to chart the ways that these coexistent bodily regimes are imbricated and experienced, especially including the ways that these create tensions and antagonisms for the subject (building from Chapter 2 to Chapter 3). In Chapters 2 and 3, I lay particular emphasis on the role of skin as a target for, and a manifestation of, coexisting bodily regimes. These two chapters are an asymmetrical pairing, showing how people can, and cannot, inhabit the uncomfortable imposition of bodily regimes differently.

The coexistence of bodily regimes can imply that they are somehow discrete, such that there cannot be movement through or across bodily regimes; more than this, that bodily regimes are somehow confined to bodies themselves – and not 'of the world'. Chapters, 4 and 5 explore the topologies and topographies of psychic space. In these chapters, we learn about the slippages between bodily regimes, as the world touches upon the psyche and as the psyche touches upon the world. Significantly, this involves the shifts and whorls of affects through the body and the social. To understand this, these chapters draw heavily on the idea of the Möbius strip. The strip is mostly understood as a way to describe the inversion of inside and outside (and is aligned therefore with the torus and the Klein bottle). However, key to these chapters is the movement along the strip that creates the inversion. It is movement that creates the tension between overdetermination and indeterminacy (at each point). Significantly, the strip also necessarily has width, which requires factoring in lateral movement – as an additional dimension of indeterminacy.

To understand the indeterminacy of bodies and affects, I focus on unconscious communication in Chapters 5 and 6 – following Paul Kingsbury and my identification of the unconscious and transference as two of the four fundamentals of psychoanalytic geography (2014, Ch. 1). These chapters take us far from an Oedipally-centred version of Freud. Instead, these chapters develop a model of Freudian thought that is focused on the ways that thought and non-thought (as in thoughts in the unconscious) are produced and repressed, but also represented, circulated and communicated. This brings us fully into a model of the subject that is radically open to the other, which necessarily begs fundamentally questions about the nature of the distribution of the senses, where no distribution – or bodily regime – can be taken as read. This then takes us (in Chapter 7) to the art of Sharon Kivland. Here, we are back on Rancière's terrain, the aesthetic. However, we are now attending to the indeterminacy of bodies, affects and politics that arises from the dynamics and disruptions of the aesthetic unconscious. This brings us full circle, back to the politics of bodies and affects that emerge from the clash between bodily regimes. However, what we now know (from the chapters within the book) is that these clashes emerge not just from the split between thought and non-thought (etc.), not solely from the coexistence of different forms of splitting (i.e. the coexistence of distributions of the sensible, of bodily regimes), but from the movement along, across and between them. With this idea, it will be possible to return to Grenfell (in the Conclusion in Chapter 8) to think once more about the interweaving of bodies, affects and politics.

The next two chapters focus on skin, both as an overdetermined location where identity, difference and experience are seemingly settled and known, but also where class, race, gender, sexuality and the relationships between them are revealed to be mutable and indeterminate.

Chapter Two
Dislocated by Epidermal Schemas: Skin, Race and a Proper Place for the Body[1]

Introduction

In the last chapter, I set out an analytical imperative that draws inspiration from Rancière's idea of the distribution of the sensible. In effect, he argues that there is an aesthetic regime that determines what can be seen and what cannot, who can speak and who cannot, who can be heard and who cannot, and that assigns bodies to a proper place. I want to dwell, throughout this book, on moments when there are tensions between different forms of determination – that is different ways that bodies can be assigned a proper place – and also when there are moments of indeterminacy. These moments are, in fact, common. Recently (as sharply revealed in the Labour Party leadership contest of 2020), for example, the Labour Party in the United Kingdom has struggled with the problem of gender and transgender. The Labour Party has traditionally supported women's groups that seek to exclude men from certain spaces, consequent of experiences of domestic abuse and sexual violence. It has also supported the rights of trans and gender fluid people to self-identify their sex and gender. As a progressive party, the Labour Party seeks to champion all those who suffer marginalisation and exclusion, especially where this is underpinned by psychological and/or physical

---

[1] Source: From Pile, S. (2011). 'Skin, race and space: the clash of bodily schemas in Frantz Fanon's Black Skins, White Masks and Nella Larsen's Passing', *Cultural Geographies*, 18(1), pp. 25–41. © 2011, SAGE Publications.

Bodies, Affects, Politics: The Clash of Bodily Regimes, First Edition. Steve Pile.
© 2021 Royal Geographical Society (with the Institute of British Geographers).
Published 2021 by John Wiley & Sons Ltd.

violence. Yet, these experiences and identities come into conflict when women born and self-identified as women wish to exclude women born as men from their spaces. The problem of the politics of identity turns around whom people really are and how we know – and both of these questions are answered (politically, socially, individually) through the body. The sharpness of these questions of identity are, to be sure, also long standing. In next two chapters, I focus on the racialisation of bodily regimes – utilising in particular the sight and site of skin.

In this chapter, I develop the idea of the distribution of the sensible as a bodily regime that not only creates and relies upon epistemic cuts through the body – eyes, mouths, ears that see, speak and hear and that do not – but also composes those body parts into a seemingly whole body that is then policed as such. Further, I show how the body-in-parts is assembled into a whole body by more than just the sense-able. Just as it includes skin and bones, it also draws in DNA and fictions of blood. It privileges specific locations in the body, such as skin and blood, while rendering other body parts meaningless. It therefore matters how bodies are parcelled up: that is, both cut into pieces and wrapped together to create a seeming whole.

Two broad arguments are made in this chapter: first, that bodily regimes are created by parcelling up bodies and by assigning them a proper place within that regime; second, that bodily regimes can coexist and overlap – and are not just the product of top-down power relationships, but also the product of lived experiences. The coexistence of bodily regimes means that the body never has just one location, and can therefore not be limited to a proper place (thought of as a singularity), rather people are constantly negotiating their dislocations: some less easily than others (see Shabazz, 2015). The politics of the body, in this vein, must draw not only upon the epistemic cuts that run through bodies to create regimes of the sense-able body, but also on the friction that occurs between the different schemas and regimes that lay claim to bodies.

Analytically, I draw on Frantz Fanon, who saw more than one way that bodies are disassembled and reassembled. In his experience, Fanon felt himself between two bodily regimes: one based on a corporeal schema; another grounded in a racist epidermal schema. I draw on these ideas to suggest that there is more than one bodily regime, where these regimes parcel up the body through a variety of schemas based, for example, on blood and skin, on voice and noise. From this perspective, the body is formed out of many schemas – utilising ideas about, for example, skin, hair, lips, noses, comportment, height, weight, blood, voice, smell and so on – each of which provides the opportunity both for particular epistemic cuts through the body and to be aggregated (however loosely or incoherently) into a bodily regime. The first part of this chapter explores Fanon's experience of the imposition of a racialised bodily regime upon him and his reaction to this imposition. Fanon observes that a racist epidermal schema splits the world into black and white. However, this schema fails against Fanon's own corporeal schema; that is, the body he is. Indeed, the racial black/white epidermal schema not only fails to delineate race, it fails as a spatial strategy to assign bodies to a proper place.

To explore this further, the main part of this chapter focuses on passing, where bodies occupy places and spaces to which they are not assigned, and being caught passing, when the consequences of policing identity are played out. Passing usually subverts the privileged marker of skin as a way to determine people's proper place within racial schemas (Price 2012). But not always. For example, in 2015, Rachel Dolezal's parents released photographs of her aged 18, showing her to be blond haired and freckled white skin (see Brubaker 2016a, 2016b). At the time, she was the president of the Spokane Chapter of the National Association for the Advancement of Colored People (in the United States) and chair of the local police ombudsman commission. Her parents declared her a fraud. Disgraced, and no longer considered black, Rachel was forced to withdraw from her roles. For many white people, she is a 'race traitor'.

Yet, she insists on her right to identify as black. Quoted in *The Guardian*, she says 'white isn't a race, it's a state of mind' (25 February 2017). More precisely, Dolezal argues that race is not about what you think or feel you are, it's what you know yourself to be. This internal truth of race does not necessarily align with social and bodily markers of race. Indeed, as skin is one of those locations where race is marked socially and physically – where the categories of white and black are authenticated and reproduced – experiences of passing reveal skin to be an indeterminate marker of race. The practice of passing underscores how skin fails to make race visible and knowable. (And not just race.) It thereby makes the use of skin as a privileged marker and maker of race and identity all the more ironic as, arguably, it is its indeterminacy that causes the policing of racialised bodily regimes to be so violent and ruthless (see McKittrick 2011).

A Proper Place for Skin? Epidermal Schemas and Their Ends

Unsurprisingly perhaps, it has been geographers of race that have had most to say about skin. Even so, mostly skin has been evoked as an aspect of the racialisation of bodies, along with other body parts, such as hair, blood, bone structure and so on. Commonly, racialised differences in skin colour are treated as if they were an effect of the signification of race in racist discourses (see Bonnett 2000). However, Kobayashi and Peake open the door to thinking about race through the body by arguing that race is contingent on biological categories (1994). From this perspective, rather than being the passive target of race discourses, bodies constantly threaten to destabilise processes of racialisation. For example, McKittrick (2000a) suggests, skin has only ever been viewed from the outside by social constructivism, leaving it inactive and inert. Instead, she argues that the relationship between race, bodies and space is volatile, not settling into neat categories of either race or space (2000b). For example, McKittrick draws on the writings of M. Nourbese Philip to illuminate the 'chaos' of black womanhood (2000a). Importantly, as Saad and Carter have argued, the body also destabilises other categorical logics, whether of gender, sex, class or whatever (2005).

So, racial locations are awkwardly occupied, at best, hinting that racist grids of meaning be avoided when discussing race and, by implication, skin – and skinly experiences. Thinking through skin, thus, means acknowledging the surface of the body, materially, psychologically and experientially: a dense surface where the social, the psychological and the fleshly are inseparable (see Gallop 1988; and, more directly on skin, see Ahmed and Stacey 2001). Importantly, it is precisely because skin is so dense a surface, so over-determined with meaning, that it also threatens to undo the very race and gender relations that are woven through it. As Minelle Mahtani has argued, the body is always already more than just a race or a class or a gender or a sex or. . .and so on (2015, especially Ch. 5).

In this context, ontologising race appears risky. Not only can this manoeuvre fix race (no scare quotes) into seemingly stable contents, concepts, capabilities and commitments (Saldanha 2006, p. 9), it can fail to acknowledge the ways in which ontologising 'race' (with scare quotes) has been itself been a highly racialised practice. Thus, Olund points out:

> In the Progressive era [1890s to 1920s in the United States], reformers sought to ontologize whiteness as mobility, agency and responsibility, and blackness as passivity, objectification and irrationality. These racially-targeted 'commitments' are needless to say quite different from Saldanha's, yet they were the twin products of Progressive era understandings of human agency as produced, as a material and mobile assemblage of objects, images, affects, practices and spaces that moved in some way to sexual equality, but maintained white privilege. (2009, p. 500)

Olund concludes that any strategy of ontologising race must also include a commitment to thinking through the relations of power that flow through such a strategy. That's fine, but what slips from view is, once again, skin. Thus, whiteness becomes ontologised as mobility, agency and responsibility, yet seemingly removed from the fact of having white skin (or not); blackness becomes passivity, objectification and irrationality, yet detached from bodies that might have black skin – whatever that is. Cauterising the social from the body not only masks the ways in which the social imposes upon the body, it also obscures the ways in which the body might construct the social and all its concepts, capacities, commitments and communities. Once more, we seem to flip flop between flawed alternatives: race as a cultural category versus race as a bundle of biological bits. To seek a way past this, I turn to a key moment in Fanon's account of the fact of blackness (Fanon 1952, Ch. 5; see also Pile 2000).

As is well known, Fanon joined the Free French Army in 1943 at the age of 18 (for a biography, see Macey 2000). He was initially deployed to Algeria. Sometime in Winter 1944 (probably), Fanon is travelling in uniform on a train. He may well have been heading north from Algeria towards Alsace-Lorraine, where he would be wounded at Colmar (and receive the *Croix de Guerre*). On this journey, famously, a small boy confronts Fanon (I will return to this moment in the

Conclusion in Chapter 8 of this book). In alarm, the boy cries: 'Look, a Negro'! Initially, Fanon sees the funny side of the boy's reaction: 'It was true. It amused me' (p. 111). But he is not amused for long. Fanon feels himself become a body: the body of a black man.

> I could no longer laugh, because I already knew that there were legends, stories, history, and above all historicity [. . .] Then, assailed at various points, the corporeal schema crumbled, its place taken by a racial epidermal schema. (p. 112)

It appears, then, that a racial epidermal schema has imposed itself upon Fanon: disassembling his body only to re-form along the lines suggested by a racialised epidermal schema. The schema is appalling:

> My body was given back to me sprawled out, distorted, recolored, clad in mourning in that white winter day. The Negro is an animal, the Negro is bad, the Negro is mean, the Negro is ugly. . .. (p. 113)

The epidermal schema is woven out of thousands of stories, anecdotes, images and so on that surround the body, giving it a kind of truth – a truth that is comprised of certainty and uncertainty: 'Mama, the nigger's going to eat me up' (p. 114). The epidermal schema, to be sure, is comprised of more than racist imagery and discourses. It is also bodily. Later in the chapter, Fanon argues that the idea of race must combine an understanding of both 'psychobiological syncretism' and 'experience', of both race and class (p. 133). The epidermal schema, then, is woven out of many threads. A proper understanding of Fanon's experience, in Fanon's own terms, cannot rely on the idea that racialised schemas are simply imposed on corporeal schemas. In his view, his experience was:

> a slow composition of my *self* as a body in the middle of a spatial and temporal world – such seems to be the schema. It does not impose itself on me; it is, rather, a definitive structuring of the self and of the world – definitive because it creates a real dialectic between my body and the world. (p. 111, emphasis in original)

Thus, Fanon implicitly sets up a 'real dialectic' between, on the one hand, a corporeal schema ('my body'), and, on the other, a racialised epidermal schema (imposed by 'the world'). The corporeal schema appears to provide the self with a place in the midst of the world, while the epidermal schema allows the body to be recomposed by the world, in his case along racist lines (see also Sullivan 2004). This needs a little untangling – as both schemas are bodily, and both racialised; further, both appear to be experiential and also psychological; and both are social. These similarities, for me, suggest that these schemas are associated with coexisting bodily regimes; regimes that are not aligned and do not add up to a single regime. There is a co-constituting tension, a real dialectic as Fanon puts it, between them.

The tension between these schemas has to do with their time and space: what appears to come from within versus what appears to come from outside. Neither exacts a space of the body that is simply internal or external: these spaces constantly switch, bleed, blur, grate, fold. What is important is not the content, the truth, of each schema, but that the tension between them acts upon both of them, making and remaking them. At stake, in this 'real dialectic', is the social relations that snare and entangle bodies. More, the friction generated by the coexistence of the corporeal and the epidermal schemas is simultaneously worldly, psychological and experiential. And racial – but only ever in part. Significantly, the implication of Fanon's analysis is that there would be more than just these two schemas, if only because (as we know) racial ontologising through the body is conducted by more than schemas associated with skin (on racialisation, see Murji and Solomos 2005).

At its simplest, for Fanon, bodies occupy two locations simultaneously: 'my body' and 'the world'. Yet, these locations cannot simply be disentangled. Rather, what Fanon attends to is the multiple ways in which bodies and worlds are co-produced, yet in ways that do not collapse one into the other – not because they are singular, but because they are already something multiple (as Saldanha argues). Following this, and as McKittrick argues, this requires a spatialisation of blackness and the black body that itself acknowledges its multiplicity, but also refuses to be collapsed onto some self-evident space of the body or of racialised schematics grounded in the body (2006, p. 27).

From Fanon, then, we can learn that there is more than one bodily regime through which the body becomes a body in the world. He identifies two, each co-constituted by the antagonism between them. On the one hand, there is a 'whole body' corporeal schema; on the other hand, there is a 'partial' racialised epidermal schema. The politics of these schemas face in opposite directions. Fanon and Olund agree that the body must be understood through its location in the racist black/white grid of signification and power (i.e. in racialised epidermal schemas). Fanon and Saldanha would seem to agree that the body must be understood through capacities and contents which are perhaps beyond, or not reducible to, social relations (i.e. through corporeal schemas).

Further, Fanon offers the possibility that we might, instead of choosing between one bodily regime and another, look instead for moments when they are in tension, and seek to understand the consequences of living through a body that is caught in more than one bodily regime (following Chapter 1). Indeed, arguably, the shock of Fanon's experience is a consequence of being caught between two unconscious distributions of the sensible that he has suddenly become all too aware of. This experience, this moment, is many sided. Fanon felt himself dislocated by a bodily regime that failed to see his body as being in its assigned, proper place. He also realised that he shared this unconscious bodily regime, which made some bodies visible, or silent, or in the wrong place. He seeks to resist the unconscious structuring of the body, so in that moment he bites back. The unconscious structuring of bodily regimes produces overdeterminations – the use of ideas to determine, many times over, the true nature and meaning of bodies – and also indeterminacies, where bodies fail to conform to these overdeterminations.

One response to indeterminacy is of course to produce a set of policing responses, over bodies and over space, designed to ensure that bodies are demarcated and stay in their proper place. The sheer energy with which societies ruthlessly and brutally police bodies and places is all too evident. However, given the focus of this book on the coexistence of bodily regimes and upon the indeterminacy of bodies, I would like to explore a moment where both policing, indeterminacy and the coexistence of bodily regimes are sharply in evidence. An opportunity is provided by the well-known practice of passing, where the boundaries that demarcate a proper place for bodies are deliberately crossed. In Nella Larsen's *Passing*, we not only witness the operation of a racist epidermal schema, but other corporeal schemas associated with skin, femininity and sexuality. As with Fanon, these schemas seem to make the body knowable, yet do not; they seem to assemble a whole body, but do not.

Passing, Races and Proper Places

In 2002, in a wide-ranging essay, David Delaney sketches out the relationship between race and space. In particular, he raises the issue of 'passing' (pp. 8–11). An analysis of passing, he argues, would tell us much about the spatialisation of race and the racialisation of space. For Delaney, passing involves an invisible, yet successful, trespassing (drawing on Ginsburg 1996; and, Kawash 1996). For passing to be possible, certain basic assumptions have to be made. *First*, space has to be segregated and bounded, with homogeneity of the population within the boundaries. *Second*, these boundaries have to be porous. *Third*, there have to be ways to fool the border guards (whatever the form that this fooling or guarding takes). Passing, as commonly understood, involves people who are legally defined as black fraudulently (*sic*) adopting white identities and inhabiting spaces marked out for white people: that is, in passing, people have crossed a 'colour line', a racial boundary, to assume a different racial identity (Ginsburg 1996, pp. 2–3). The motivation for passing, in this context, is clearly both to escape racial oppression or discrimination and also to claim the privileges and status of the dominant race.

Unsurprisingly, for Delaney, the idea of passing, as such, is primarily derived from US history and its discourses of racial difference. It is, consequently, strongly associated with African-American literature (see Bennett 1996; or Wald 2000). In this context, to pass as white requires more than simply the cultural resources to assume a white identity, it also requires a skin colour that permits an individual to look visibly white to others. Passing, then, involves not only a cultural identity, it directly provokes a question about the body itself – what does it mean to have a skin that can pass? Or, put another way, what can skin do (or not)?

Clearly, passing is a spatial practice. It is not simply the adoption of another cultural identity, it also means crossing a line, marked in space. The practice of passing can, therefore, involve people crossing in other directions, across other seemingly clear-cut and spatially demarcated forms of identity (see Sánchez and Schlosberg 2001). For example, white men have passed as black men (Griffin 1961; Wald 2000, ch. 5; and, Lott 1993); women pass as men in seemingly male-only domains, particularly (but not only) the military (see Creighton and Norling 1996): as soldiers in the American Civil War (and probably every other armed conflict before and since) (see Blanton and Cook 2002); as sailors aboard pirate ships and warships (Stark 1996). Practices of passing, then, are variable historically, yet there are some similarities. They involve: some form of disguise or pretence; inhabiting a space that is, in some way, exclusive; and, consequently, crossing a cultural and spatial boundary, which is policed – yet not well enough.

It is the ability to fool the border guards of identity, and to call into question the nature of identity, whether founded culturally or corporeally, that gives passing its radical or transgressive political edge (Ginsburg 1996, pp. 3–5; Ahmed 2000, pp. 125–133). The basic problem is this. *On the one hand*, passing renders identity unknowable, culturally or corporeally, and this consequently makes the policing of space arbitrary. Passing should, therefore, destabilise all attempts to secure pure identities and to purify space for those identities. Yet, *on the other hand*, passing also fixes identities: those who pass do not, ultimately, challenge or destabilise cultural and spatial boundaries, but leave them intact. Thus, the practice of passing is politically ambiguous: both fixing and transgressive of identity and space (Ahmed 2000, p. 130).

For Delaney, the fact of passing as a lived experience can nonetheless tell us much about the 'complex notions of deceit, betrayal, suspicion, and anxieties concerning threats to racial purity' and 'it may be of value in coming to understand imaginative geographies' of racial formation (2002, p. 9). Passing, for him, is primarily a racial experience: to pass oneself off as belonging to another race; to be suspicious of the race of those around you; to escape racial oppression by crossing the race line. There is a clear geography to this: of segregated spaces, either white or black; of boundaries between spaces; and, of crossing over. Further, passing in the racial space of the United States also evokes spaces of the body: *on the one side*, there are the racial demarcations based upon invisible markers, such as, crucially, blood and heritage; *on the other*, there are the visible markers of race, such as hair, skin, bone structure and the like. To pass, in this sense, is less about disguise than about assumption: what people assume about the body and how people assume identity. Certain bodies can 'assume' certain identities, while others cannot.

Perry Carter (2006) has exposed this issue in his insightful reading of Nella Larsen's 1929 short novel *Passing*. In his view, passing is fundamentally about:

> how certain bodily ambiguous individuals, undecidables, have a *choice* as to which racialized roles they wish to perform. Certain individuals are able to make this

choice because identities – whether based on 'race,' gender, sexual orientation or class – are not fixed but fluid. Yet often space acts as a fixing agent, binding identities to certain places. Consequently, when these subjects deracinate and transport themselves, their identities become unstable, allowing them, if they so choose, to perform entirely different identities. (p. 228)

In this view, passing demonstrates that identities and bodies are mutable, indecipherable, mobile and changeable. If there are fixities at all, these are spatial. These spatial fixes to the problem of racial indecipherability and mobility lie in the creation of boundaries and in guarding the homogeneity of space. Bodies and space are, in this view, social constructs, built out of invidious and injurious power relations – in this case, between 'races' (the use of scare quotes being *essential*). While Carter's article sets out a different challenge than Delaney's, they are intimately linked: mobile (racialised) bodies threaten solid (racialised) spaces. But, if race really is so mutable and indeterminate, then there is a question as to how it is possible that racialised spaces can act to fix or stabilise race.

One answer is to see the racialised body as an effect of language or culture: thus, space and race are co-produced, symbolically, as fixed and determinate. This is the line taken, famously, by Judith Butler (1993, especially ch. 6). For her, Nella Larsen's novel is a question of 'what can and cannot be spoken, what can and cannot be publically exposed' (p. 169). Thus, for Butler, skin becomes a 'sign' of the symbolic order, which she suggests is phallic (white, privileged and heterosexual) (see pp. 181–185). In this analysis (following Lacan), the adoption of a position within the symbolic order represents 'a living death' (p. 185). Thus, for Butler, Irene's survival is actually a symbolic death. Saldanha (2006) disagrees: bodies cannot, and should not, be decorporealised by evaporating them into the symbolic (p. 12).

For Saldanha, the answer to this puzzle is that bodies already have race (no scare quotes): race, here, is not singular, but rather multiple. Bodies have races, not a race. Thus, bodies threaten spaces not because they are mutable or indeterminate (in language and culture), but precisely because bodies are already something – a phenotype, in Saldanha's terms – prior to their immersion in processes that construct them as being of a race, culturally and corporeally. Central to this dispute, and also to racialised narratives of passing, is above all visible difference – and not just any visible difference: in dispute is *skin*. To be sure, skin is not the only marker of race, but it is skin that makes phenotype self-evident and confirms its ontology, rather than say hair, or bone, or voice or blood (see also Saldahna 2010).

For me, a focus on skin itself provides a different, and unsettling, angle on the question of the relationship between racialised bodies and racialised spaces. This is because skin refuses to settle into binary logics that code both bodies and spaces as black or white. Yet skin also resolutely refuses to be colourless: despite protestations that race is socially constructed, skin determinedly asks that bodies be recognised as having a race, even where these are many and tiny. Is this a paradox or a contradiction? In this chapter, I use Frantz Fanon's notions of epidermal and corporeal schemas to think through this question – as it was also a question for him (especially Fanon 1952, see Ch. 5).

I believe that Nella Larsen's novel *Passing* can also be read as an attempt to grapple with the problem of skin (see Davis 1994). Her account of skin in some ways matches Fanon's. They both celebrate blackness, while also understanding that blackness is not a fact as such. They both think relationally about how the body is racialised and also about the politics of racialisation. For both, the world is experienced through skin. Skin: a semi-permeable layer that lies between interior and exterior worlds, constantly crossed (see also Pile 2009; see also Chapter 6). Something of this can be seen in the epigraph that Carter cites at the top of his essay:

> We encounter the world in our bodies, and through our bodies' most exquisitely sensitive sense, our skins, we take the world into ourselves. We have made and remade a world where nearly every experience is shaded and shaped by the color of those bodies, the tones of those skins. (Lazarre 1997, p. 94; cited by Carter 2006, p. 227)

Skin shapes the world – yet not freely so: as it projects its colour into the world, as experience itself is coloured by it. This is a hard argument to get right: it raises the spectre of biological determinism that geographers, and social scientists, have long been wary of (as Saldanha shows, 2006; 2010). Worse, giving agency to bodies, to skin itself, seems to deprive politics of a means of intervening in race and racialisation. Perhaps, however, by starting with skin, rather than bodies or processes of racialisation, we might find another way to address the problem.

So, I would like to puzzle for a bit about a skinly view of the world be re-examining Nella Larsen's *Passing*. The novel explores the shifting and mobile lives of two African-American women: one, Irene, occasionally passes as white; the other, Clare, has chosen to pretend to be white. The characters grapple with the consequences of these choices. The novel, then, presents the reader with the personal, and political, desires and dilemmas of these women – not simply as fictional characters, but as emblematic of the desires and dilemmas of black women seeking a place in a white/male dominated, spatially segregated, society (Kanneh 1998). This work has drawn a vast amount of critical comment, yet curiously the role skin plays within the novel has often been taken for granted. Perhaps it is because skin is clearly what is at stake in the story that it is almost too obvious a dimension of the story to warrant analysis in itself. Rather than skin, it is the expression of femininity and sexuality that seem to refract the experience of race (especially when contrasted with Fanon's own, male, experience: Kanneh 1998, pp. 174–178). Yet, attending to Larsen's sensitivity to skin and colour yields insights into how she conceptualised race and space. More than this, however, thinking through the friction between competing epidermal schemas in *Passing* troubles what it might mean to have or be 'a race'.

Colour Sensitivity

The narrative drive in Nella Larsen's *Passing* revolves around the intense relationship between Irene Redfield, on the one hand, and Clare Bellew (née Kendry), on the other. Irene Redfield, though she could pass for white, only does so pragmatically and temporarily. Clare Kendry, meanwhile, is living a white life, though she is – by the racist 'one drop' rule – in fact black. Both are married and middle class. Irene is married to a doctor, Brian, who is working in the poor neighbourhoods of Manhattan. Clare's spouse, John (Jack), is an international banking agent, which involves much travel to major cities in America and Europe. Both grow up in Chicago, but when Clare is 12, her mother dies so Clare's white aunts take her in. Irene grows up black; Clare grows up white. Separated, they meet by chance in the whites-only tea room of Chicago's Drayton Hotel (perhaps modelled on the Fountain Court at The Drake Hotel on Chicago's Gold Coast). At this point, both Irene and Clare are passing, and their initial contact is fraught with the anxious possibilities of exposure, recognition and misrecognition, and the delight and awkwardness of seeing one another.

Sometime later, Irene receives a letter from Clare. By this time, Irene has moved to Manhattan. Clare, it turns out, is also in New York. Clare wants to meet. Despite herself, Irene agrees, and instantly regrets it. Over the coming weeks, Irene and Clare's lives entangle further and deeper. At each turn, Irene feels unable to prevent Clare's ever greater penetration of her life. Until, at last, Irene kills Clare by pushing her out of a window. Larsen never makes this denouement explicit, leaving the reader to draw this conclusion, having eliminated the other possibilities – that Clare jumped, that she fell, that she was pushed by someone else (with Jack Bellew being the prime suspect). Of course, the novel can be read as one in a long line of 'tragic mulatto' stories. Yet, the novel has drawn persistent and on-going alternative readings.

As the novel is about passing – a deception of identity, bodies and space – other hidden metaphorical layers have been excavated (see Kaplan 2007). In Tate's (1982) understanding, the book is more like a melodrama, involving the tensions and failures resulting from the dark secrets within marriage (see also Tate 1992). In a reading that reignited interest, McDowell (1986) argues that *Passing* is a work of lesbian fiction that *passes* as straight fiction. As the novel has been taken as emblematic of real experiences of race, femininity and sexuality, these readings have become recursive, folding back onto Nella Larsen's life, which has itself been read as a 'tragic mulatto' story (Mafe 2008). Significantly, Carby (1987) argues that the novel represents one of the earliest attempts by an Afro-American woman to find a place for black women in the city. In particular, Irene and Clare

are seeking a place within Chicago and Harlem in New York. Harlem, especially, holds significance within African-American history: its importance stretching far beyond 125th Street. Harlem, Taylor observes:

> stretches beyond the confines of its physical space to gather in a far-flung African diaspora. The distinctiveness of African, Caribbean, and southern and midwestern U.S. blackness abounds in the sights, sounds, colors, and flavors of the traditions and day-to-day life in the community. (2002, p. xv)

Although a long-standing racialised ghetto, by the 1920s, Harlem had become home to an aspiring African American élite. It was also a magnet for a wide-range of African American artists and intellectuals, including Langston Hughes, Marcus Garvey, Duke Ellington, Josephine Baker, Bessie Smith, Paul Robeson and Nella Larsen amongst many others (see Baker 1987; or Powell 1997). Many of these artists and intellectuals – including Nella Larsen – used Harlem as context to reflect upon the experiences, and future, of African Americans. In Harlem, 'black artists and intellectuals participated jointly in the creation of a new urban collective identity' (Taylor 2002, p. 8) – famously known as the Harlem Renaissance. As Alain Locke, one of the key figures of the Harlem Renaissance, put it:

> In Harlem, Negro life is seizing upon its first chances for group expression and self-determination. It is – or promises at least to be – a race capital. [. . .] Without pretense to their political significance, Harlem has the same role to play for the 'New Negro' as Dublin has had for the New Ireland or Prague for the New Czechoslovakia. (1925a, p. 7)

Harlem was not simply a postcolonial capital city for African Americans within the United States, it was also meant to be home to a 'new Negro'. The 'new Negro' was to be a 'gold standard' against which African Americans would measure their progress towards a postcolonial subjectivity: it was both a political demand and a way of life. Or, as Jack Bellow puts it: 'Great city, New York. The city of the future' (*Passing*, p. 173).

For Dawahare (2006), Larsen's work inscribes the 'gold standard' of proper black behaviour demanded by the Harlem Renaissance (for an anthology of original writings by members of the Harlem Renaissance, see Locke 1925b). In this reading, Larsen's *Passing* is about the internal tensions within debates about black class advancement. Taking these readings together, *Passing* articulates desire; female, lesbian, black, class, urban. Desire, yes; but, also its contradictions. There is no easy desire here; nor, consequently, is there an easy politics that might flow from it – and reading the book for its colour, its skins, will underscore this point.

In her perceptive discussion of the colour in *Passing*, Jennifer DeVere Brody (1992) shows that Larsen presents African Americans in a spectrum of colours.

Brody insists on the importance of colour in Larsen's novel. Despite this, she reads skin colour against the 'true' race of the characters, thereby ironically undermining the significance of skin colour in favour of other racialised makers (and indeed class position). For Dawahare, on the other hand, Larsen's epidermal schema equates lighter skin with greater social value (2006, p. 33). While admitting that the spectrum of colour does not exactly map onto racialised hierarchies, he insists that the spectrum of colour largely corresponds with racialised values that have poles defined by ugliness and beauty, and by the semibarbaric and sophisticated. Whiteness, meanwhile, is always associated with money and privilege (similarly, see Dwyer and Jones 2000).

For me, it's worth another look at how Larsen paints skin colour and uses colour more widely: partly because the anchor point of whiteness is not quite as fixed and stable as it at first appears, partly because of this curious failure to be able to map colour onto race onto privilege onto money – precisely because, I argue, there are different epidermal schemas in play.

Passing is saturated with colour: Larsen is constantly detailing the colour of things. For example, on the first page, on reading Clare's purple-inked letter, Irene recalls 'a pale small girl, sitting on a ragged blue sofa, sewing pieces of red cloth together' (p. 143). Pretty much everyone has black/dark eyes and bright/black hair. Yet, Clare's hair is golden. Golden is a particularly significant colour: on the significance of golden skin, see Cheng 2010. It is Larsen's colour sensitivity to skin that intrigues me.

On the Redfield side, Irene is 'warm olive' (p. 183 and p. 145; allowing her to pass for Italian, Spanish, Mexican or Gypsy, p. 150), 'beige' (p. 183, at least her foot is) and 'dark white' (p. 218); her husband, Brian, is 'deep copper' (p. 184; which means he cannot pass, p. 168) with 'tea-colored fingers' (p. 186), and their son, Junior, is similarly toned (p. 192); Irene and Brian's maid, Zulena, is 'mahogany' (p. 184); a servant, Liza, is 'ebony' (p. 164).

On the Bellew side, Clare is 'pale' (p. 143), 'ivory' (pp. 148, 161, 198 and 220, allowing her to pass for white) and 'white as a lily' (p. 171); her husband, Jack, has 'an unhealthy-looking dough-colored face' (p. 170); Clare's two aunts are 'white' (p. 158).

In their social world, a mutual friend, Gertrude Martin, has a 'large white face' (allowing her to pass) and is married to Fred, 'a white man', who knows that 'she [is] a Negro' (all p. 165); at the Negro Welfare League's dance, there are white and black men, pink and golden women (p. 204); amongst Irene's (and later Clare's) friends, Hugh Wentworth is 'Nordic' (p. 205), Ralph Hazleton 'unusually dark' (p. 205) and Felise Freeland 'golden' (p. 226).

To say the least, Larsen presents a rich spectrum of skin colours. As Kanneh points out, there is an almost baffling array of shades, complexions and textures to skin (Kanneh 1998, p. 175). This is evident, also, in Larsen's other major work, *Quicksand* (1928). For example, its main protagonist, Helga Crane, sits watching the Harlem [New York] scene:

For the hundredth time she marveled at the gradations within this oppressed race of hers. A dozen shades slid by. There was sooty black, shiny black, taupe, mahogany, bronze, copper, gold, orange, yellow, peach, ivory, pinky white, pastry white. There was yellow hair, brown hair, black hair; straight hair, straightened hair, curly hair, crinkly hair, woolly hair. She saw black eyes in white faces, brown eyes in yellow faces, gray eyes in brown faces, blue eyes in tan faces. Africa, Europe, perhaps with a pinch of Asia, in a fantastic motley of ugliness and beauty, semi-barbaric, sophisticated, exotic, were here. But she was blind to its charm, purposely aloof and a little contemptuous, and soon her interest in the moving mosaic waned. (pp. 59–60)

Helga is curiously detached from the race of races through Harlem's streets. It is as if she has become overwhelmed by, or lost in, the sheer volume of epidermal and corporeal schemas that she is presented with. This, however, is not the situation that Irene and Clare find themselves – they struggle with the epidermal schemas that (seek to) define them. More than this, like Fanon, this struggle plays itself psychologically, through the idealisation and denigration of the skin they are in. On the one hand, Fanon affirms: 'I am a Negro, I am a Negro, I am a Negro. . .' (1952, p. 138). On the other, he despairs, 'as I begin to recognise that the Negro is the symbol of sin, I catch myself hating the Negro' (1952, p. 197). The (political, personal) question that is presented by Fanon, that also confronts Larsen's characters, is this: is skin to be celebrated or denied? Answering this question, in practice, depends on how bodily regimes – with their defining epidermal schemas – compose bodies, in the moment, in a particular place.

Epidermal Schemas in *Passing*: Monochrome versus Full Colour

It is important to note that (at least) two kinds of bodily regime run through *Passing*. On the one hand, there is the abstract epidermal schema associated with wealth and power, with the segregation of races and spaces (and it is through this that *Passing* is mostly read). On the other hand, there is the kaleidoscope of skin colours: an epidermal schema that can be seen in a spectrum that ranges from mahogany and ebony, through copper and golden, through olive and pink, to dough and ivory. Both are colour-sensitive, both 'make up' the lived experience of skin, albeit it differently so. Nonetheless, the black/white epidermal schema, by itself, cannot provide an account of the lived experience of skin (see Saldanha 2006). Nor can this schema account for, or even corral, the sheer volume of colour Larsen uses in her novel. Dawahare was, therefore, right to be suspicious of mapping darkness and lightness onto lesser and greater value, respectively. There are at least three consequences of the failure of the black/white epidermal schema to map directly onto Larsen's colour-rich epidermal schema.

First, surface colour does not correspond to the truth of race identity. Of course, we know that Clare is black passing for white, but Larsen's truth of racial identity is more a question of where ravishing beauty and the exotic lie (see especially *Passing*, pp. 205–206). In Clare, who has white skin, what is exotic lies *below* the skin, while the unusually black Ralph Hazleton's handsomeness lies *in* the colour of his skin. *Both* Clare and Ralph are considered (by many) to be objects of intense desire: in Ralph's case, because of his blackness; in Clare's case, her invisible blackness actually adds to her attractiveness, even while her self-evident whiteness is itself seductive and desirable. Inner and outer blackness are, for Larsen, both seductively beautiful – though not, it must be noted, for Irene and Jack.

Second, colour changes according to personal truths, in *Passing* this is especially evident when people blush. Early in the story, 'brilliant red patches [flame] in Irene's warm olive cheeks' (p. 145): indeed, she repeatedly blushes (pp. 154 and 167). Clare seems, at first, capable of repressing her emotions, and of using her whiteness as a cold mask, yet redness appears in her cheeks (pp. 169 and 198). Each blushing is a tell-tale sign – when Irene feels a mix of embarrassment and anger, when Clare feels exposed and vulnerable. Blushing is a fissure, where the disjuncture between personal truths and social masks become emotional. Here, skin is shown to be both disguise and traitor, both lie and truth. More, it shows that skin is associated with other colours, in this case blood red. Skin is, to be sure, about more than race (as explored by Benthien 2002; and, Connor 2003).

Blood, of course, is highly racialised. Larsen herself makes the 'one drop' rule bizarre and arbitrary by referring to it in pseudo-scientific percentage terms (p. 171): even so, it persists, one drop of so-called black blood still makes someone black in the United States (witness the racialisation of Barack Obama). In effect, blood determines someone's actual skin colour (and not vice versa). For example, Clare is 50% white, but by blood 100% black, though she looks 100% white. Yet, blood also shows that skin is not known. In saying this, I am not arguing here that Clare's race is undecidable either because the colour of her blood is not visibly black or because her skin is visibly white. It is, rather, that her skin is a variable surface that sometimes reveals and sometimes conceals her colour and her character. It is only in moments of blushing that her truth – her concealed identity, her true feelings – threaten to break surface. In fact, Clare's problem is that she is *too decidable*: she is always over-determined by the assumptions and meanings of others.

Third, skin can change over time. The clearest example of this in *Passing* is the cause of Jack's nickname for Clare Kendry: 'Nig' (pp. 170–171). Listen to Jack explain:

Well, you see, it's like this. When we were first married, she was as – as – well as white as a lily. But I declare she's getting darker and darker. I tell her if she don't look out, she'll wake up one of these days and find she's turned into a nigger. (p. 171)

Though self-evidently racist and abusive, sensitive Irene laughs out loud. The joke that he is unwittingly surrounded by black folk trumps her anger at Jack's overt racism (for a discussion of this joke, see Ahmed 1999). In most ways, Larsen is not saying that Clare's skin is actually changing colour. More likely, she's arguing that Clare cannot escape her true colour and that this will become increasingly difficult, even while white folk ignore what is right in front of their eyes. Yet Clare's becoming black points to the mutability of social categories of race, of true colour. Thus, there is a darkening and lightening of skin over time. Skin is also a map of personal experience, of the memory, however it is an uncertain one: perhaps revealing, perhaps deceitful. Similarly, Irene also changes colour. She is 'olive', 'warm olive', 'beige' and 'dark white', almost as if she were a chameleon whose skin changes according to context; a context that is simultaneously social and emotional.

The underlay of epidermal schemas in Larsen's work reveals some of the difficulties of simply arguing that race should be reontologised. To begin with, race has no ontology except through its schemas, even if these schemas are markedly different from the racist black/white epidermal schemas that dominate race thinking. Race is simply too mutable to 'be': its depth is shallow; it cannot be *seen* in its proxy, yet defining, corporeal location, blood; and it most definitely, definitively, changes over time. Put another way, the production of the racial phenotype, even in its guise as a thousand tiny races, is still a response to the unknowability of race. From Larsen and Fanon, however, we can see that the mutability of pluralistic corporeal schemas is precisely their advantage over black/white epidermal schemas – *and their disadvantage*. There is friction between the epidermal schemas presented by Larsen and Fanon. This takes us to Clare's colourful white skin.

Clare's White Skin: Epidermal Schemas, Race and Space

Larsen describes Clare's skin as 'ivory', but she means very different things by this. At first (p. 148), ivory skin signifies a sexualised beauty in, what is for Irene at that point, a total stranger. Clare is attractive, with almost black eyes, yet it is the wide mouth, with its scarlet lips set in ivory skin, which indicates that her beauty is sexual (see Irigaray 1977). Ivory, at this point, connotes a classical form of beauty, as if Clare had been sculpted out of marble. Of course, by using ivory as a metaphor (rather than marble), Larsen hints that Clare's source as African. Yet, she is also removed from this source, metaphorically bone removed from the body.

Like her skin, Clare is detached, dislocated. This aloofness, abstraction, itself becomes a form for passion. On page 161, we read about the 'tempting mouth', only this time the 'ivory white' has a 'soft lustre'. She is eye-catching, luminous: her skin shines with her sexuality, but she is now fleshly, skinly. This skin is not now the skin of a sculpture, but the skin of a seductive woman (see also p. 221). Clare's ivory skin is blushing on page 198. Here, ivory skin is not just a surface, capable of being written on (in this case by her emotions), but it is also (therefore) a site of vulnerability. Gone is the 'detached beauty' and 'human sexuality', to be replaced by anxiety and transparency. Yet, by page 220, the ivory skin has become a mask: a surface detached from the underneath, the true anxious space of the emotional body. It is this connotation of 'ivory skin' that has attracted almost all the critical attention (see, e.g., Brody 1992). Whiteness, in this view, is unmarked and powerful, as it is a disguise, a trick, a mask.

Indeed, whiteness and white skin are commonly interpreted as 'unmarked'. Key to this reading of whiteness is Richard Dyer's *White* (1997). In fact, reading Clare's whiteness/white skin suggests the exact opposite. Her skin signifies: beauty, vulnerability, sexuality, anxiety, detachment, deceitfulness, humanness and so on. This suggests that one-dimensional readings of whiteness and white skin have missed the point. Clare's skin changes through the book, depending on the social setting, depending on her emotional state, depending on what the author wishes to say about her character. There is no single, unmarked reading of white here; whiteness is every bit as marked as blackness. Clare is marked by her white skin; her white skin is itself remarkable and marks her out.

Even while Clare's skin changes over time, and in different social spaces, it is important to note that skin also produces social spaces. Brody notices:

> the spatial and ideological positions held by Clare and Irene are revealed in several scenes in which they interact. These scenes occur in the most 'civilised' of places – tea rooms, parlours, boudoirs and ballrooms. (1992, p. 1056)

It's also worth noting Larsen's novel exposes a deeply racialised urban geography, contrasting poor working class and well-off middle class black urban experiences in Chicago and New York. Irene, herself, feels part of the Harlem Renaissance, thoroughly at home in the aspirant rising towers of late 1920s New York (see Taylor 2002; and also Skipworth 1997; and Pile 1999). Arguably, Larsen maps her epidermal schemas onto the city and its spaces. We find certain spaces dominated, for example, by racist monochromatic epidermal schemas: the white suburbs, segregated restaurants and hotels, societies promoting black advancement. Yet, other spaces appear dominated by the colourful cosmopolitan epidermal schemas: the streets, transport hubs and, perhaps surprisingly, domestic spaces. That said, it would be better to say that Larsen shows how these schemas coexist and interact, though not necessarily (or, rather, rarely) as equal partners, as we will see.

Larsen's novels are not simply about black women finding a place in the city, for their worlds extend far beyond the city – and beyond America. Clare, for example, has travelled around Europe. While Irene hasn't, the world Irene inhabits extends (at least) to Africa, Latin America and Europe: she traces her lineage to Africa; her husband fantasises about working in Brazil; through Clare, she hears about life in various European cities. This is an already globalised world, through which African Americans, amongst others, circulate. Yet this globalised world is constantly posing questions. For these women, it is less about where they are from, nor where they are, but where they are going. Africa or New York or Brazil or Chicago or Harlem do not self-evidently mean something as their meaning is constantly being struggled over – and so do not necessarily provide an answer to these questions. Geography, at whatever scale, like skin, does not provide a ground for the truth of identity, space or race. This point chimes with McKittrick's discussion of Fanon's geographies (2006, pp. 24–25). Instead, the locations in *Passing* ask of Irene and Brian and Clare, what do these places mean to you?

Skin, meanwhile, makes spaces. Perhaps the clearest example of this is the 'sitting room' in the Morgan hotel. The Morgan is probably Chicago's Majestic Hotel, near Lincoln Park, then newly built; its sophisticated sitting room situated near the lobby. It is in this room that Clare, Irene and Gertrude meet for tea (*Passing*, pp. 164–176). When joined by Jack, it is in this room that his racism and charm become uncomfortably, and laughably, expressed in appalling detail (see above). Not unreasonably, the hotel room can be read as a white public space that enables the policing, and over-determination of the meaning of, the bodies it contains. Carter, for example, concludes his analysis thus: 'places – spaces embedded in systems of meaning, value, and power – reflect themselves upon the bodies of those inhabiting them' (2006, p. 238). For him, racially undecidable Clare, Irene and Gertrude are fixed as white by the space of the hotel room; an identity and a place that only the ambiguous physical appearance of their skin allows them to choose. This is correct. In such an analysis, the colour of race and space are given and known by the racist epidermal schema that produces them. However, we know that skin is, for Larsen (and Fanon), indeterminate and multiple. That is, that neither race nor space are the product of one epidermal schema, so other ways of understanding colour and the body can begin to interfere. Indeed, this is why Jack's comments become so awkward. Thus, the sitting room itself is an awkward space, an awkward mesh of colours and styles.

Here is Larsen's description of the room:

> Entering, Irene found herself in a sitting-room, large and high, at whose windows hung startling blue draperies which triumphantly dragged attention from the gloomy chocolate-colored furniture. And Clare was wearing a thin floating dress of the same shade of blue, which suited her and the rather difficult room to perfection. (p. 165)

The difficult room was made perfect by Clare's 'outer skin', her choice of clothing. The ostensibly white women work to make the room comfortable and easy. Their demeanour, their civilised tea, their conversation, all work to domesticate – to make domestic – the room. This sitting room is no longer simply a white public space that imposes its systems of meaning, value and power on them; they impose themselves, their skins, upon the room. It is no more a white room than a black room; in fact, it is more a black room than a white room. Yet, this is only to consider the racialisation of this space through one bodily regime, grounded in a racist monochromatic epidermal schema. Larsen's point is that the room is also racialised through an epidermal schema that is rich in colour. The room itself isn't even black or white. Gloomy it may be, but it is also colour full. It is, in part, its colour fullness that enables the women to make the room their own.

The room, occupied by the women, is more a private space than a public space. And this is precisely what enables Irene to laugh so long and so hard at Jack's failure to recognise that he is surrounded by black women; he has also walked into a black space where the women feel comfortable, and he – definitely *he* – should not. They are at home; he is not – albeit unwittingly so. This is not to undermine Carter's point: Irene's laughter is cut short, when she realises that it might expose Clare's 'true color' to Jack. But the point here is that it isn't entirely helpful to think about bodies or spaces as producing or imposing themselves *completely* on one another.

There is a real friction between bodies and spaces. It is both as thin and light as Clare's dress, and also as dangerous and as strong as Irene's laugh. Both Clare and Irene are defined by their 'second skin', their clothing, their comportment, yet they antagonistically come to define one another. Clare's superficial epidermal white schema clashes with Irene's deep corporeal black schema (to use Fanon's terms). This friction destabilises any simple segregation between a black/white epidermal schema and a colourful epidermal schema. Clare and Irene inhabit both schemas, as they see one another, as they assume others see them, and as they see themselves. Clare is white, ivory and black; Irene is black, olive and white. Yet, Clare will side with the epidermal schema that allows her to be white; Irene will side with the epidermal schema that insists Clare is black. This conflict proves deadly. Irene kills Clare.

By the end of *Passing*, the ideological choice – to pass or not to pass? – between Irene and Clare has turned to melodrama. Irene pushes Clare out of a window, and she falls to her death. Yes, others have argued that the ending is ambiguous because Larsen never explicitly says what happened, and even provides alternative possibilities for Clare's fall (see Carter 2006). Nonetheless, we must conclude that Irene acts to save her marriage to Brian by ending Clare's life. Clearly, this family melodrama is also a racialised social drama, but it is also a bodily drama.

In killing Clare, literally, *Irene saves her skin*. Sure, in one sense, Irene lies about what she knows about Clare's demise to save herself. But it is also the meaning of being black that gets saved. That is, Clare's death resolves the intensifying ideological choice between her and Irene. This choice has race itself as its stake. Clare threatens not only race as an ideal, but also race as a body. Put another way, Clare's location on the white side of the monochromatic epidermal schema threatens to undermine Irene's location on the black side, as Clare's movement from one side to the other reveals both the indeterminacy of the two sides and also of the permeability of the boundary that seeks to separate them. That is, Clare makes Irene unbearably aware of her own dislocation by monochromatic and colour-rich epidermal schemas. This, Irene cannot tolerate.

In part, when Clare dies, Irene saves her race as a political moment and also as a culturally coherent production, in which visible differences and similarities make sense. But Irene also shores up her personal sense of her body as black – and her relationship with her husband as a black family, in a progressive black neighbourhood, in global city where black people have a place. Thus, Irene also saves a sense of the body as having a race that is both identifiable and coherent, that is bounded, and that also has capacities to do things. More than this, Irene also shores up the sense of where black people and white people are supposed to be: disguising one's identity to live in the 'wrong' space is, for her, intolerable. Irene simply cannot live with Clare's destabilisation of the boundaries between white and black spaces: both by inhabiting them, and by crossing them. In the novel, Clare's and Irene's epidermal schemas clash ideologically, experientially and spatially – fatally.

There is no reason to abandon the accepted readings that the novel explores themes of race, sexuality, class and so on. Skin, however, tells us that this novel should not be read simply through dichotomised black/white (or class-bound; or sexual) categories of meaning, value and power, as this overwrites its copper, ivory, lily, golden shades of experience. And this is political: Irene's black and white politics of identity versus Clare's selfish operationalisation of the capacities and potentialities of her own body.

Conclusion: Dislocated by Bodily Regimes

I have hinted that interpretations of passing – which present a particular geography of skin – install a specific geography of race: a black and white world, with vigilant and cruel, yet ignorant and flawed, border guards. I have argued that this presentation of a black and white world catches, very accurately, an abstract system of signification and power as it is applied to bodies and space. For sure, this racist and racialised schema circumscribes people's experiences of space and place. It is commonly argued, however, that actual bodies escape the epidermal black/white schema. Indeed, some would rather start with bodies, rather than with the epidermal schemas that define them. This presents us with a problem: either we start with an epidermal schema, which defines bodies in terms of monochromatic systems of meaning and power; or, alternatively, we start with alternative bodily schemas, which define bodies through their (pre-social, aggregated) racial phenotypes (drawing on the work of Arun Saldanha).

Drawing on Fanon, however, I have sought to see these epidermal schemas, not as fixed and coherent 'end points', but relationally, where both (and more) can be 'in play'. To understand how this distinction between different kinds of corporeal schemas might play out, I examined the use of epidermal schemas in Nella Larsen's novel *Passing*. Anticipating Fanon, Larsen plays out a friction between her leading protagonists, with Irene representing an epidermal schema that privileges black skin as a social and spatial location, and Clare representing an epidermal schema where skin, space and identity are mutable and traversable. This is not a creative friction that leads, ultimately, to a cosy social, political and spatial resolution. The schemas brutally clash. But, from this clash, it is possible to surmise the presence and possibilities of bodily regimes: epidermal, corporeal, colour-rich or otherwise. Thus, for example, even the colour white (so long considered invisible, unmarked, unremarkable) turned out to have more than one meaning. Nor did it have to have *a* space.

This chapter's focus has not been on 'the total body' (to borrow Lefebvre's phrase, 1974), either as an abstraction or as a lived location. Its focus has been on skin. This has been to better see the indeterminacy and multiplicity of the schemas that seek to define the race of the body or bodies – and their spaces. In *Passing*, we saw that Nella Larsen's colour sensitivity illuminated not only the varieties of skinly experiences, but also how skin made those experiences and produced spaces and places in its own light. Of course, this observation supports the idea of the spatialisation of race and the racialisation of space, but there is also evidence that there is more than one schema that racialises and spatialises, through the body. Arguably, there are never less than two. There are, conceivably, many. Thus, though racialisation and spatialisation can produce black and white outcomes, there are always likely to be other outcomes – however hidden or marginalised. Paying attention to skin has enabled this to be made visible. More than this, skin itself reveals where the epistemic cuts lie that hide the body behind its appearance(s).

That said, there is far more to skin than its colour. It sweats. It blushes. It touches. It's a surface of pleasure and pain (as we see in the next chapter). It's vulnerable, yet resilient: a map of injuries and healing. It stretches and sags. It experiences. It remembers. It changes. Avoiding early declarations that skin is about its social construction as a surface of, or at the intersection of, race or class or gender or sexuality or. . .and so on. . .enables us to see that there're many things going on in, and emergent from, the body that cannot be simply put down to systems of meaning and power. Not so simply, at least. Irene's saving of her skin can, of course, be read through grids of superiority and privilege – and their subversion. On the other hand, betrayal, shame, guilt, desire may come from places that cannot be reduced solely to their supposed location within latitudes and longitudes of meaning and power.

It would be reassuring to feel that a fluid and multiple understanding of space and place might enable a progressive politics of identity, where markers of identity, such as skin, would be unimportant. Yet the idea of erasing skin ought to be haunted by the figure of Clare, whose skin was erased to save someone else's. Maybe Irene could have saved her skin another way? Perhaps, then, we might wish to think again about the skinly world, as Fanon urged, through its various corporeal and epidermal schemas; indeed, following Larsen, to think about making the world *even more colourful*. Like space, like place, skin is never just one thing, nor does it make just the one geography.

For both Fanon and Larson, skin is a site and sight where bodily regimes operationalise their policing of bodies. For both, this policing is brutal. Symbolically, at least, murdering the body. And, historically, of course, actually murdering bodies. Policing bodies has become no less ruthless: the policing of blackness and whiteness remains fully operational. Operating, primarily, through skin. It might be hoped that the indeterminacy of skin – its failure to reveal or authenticate who or what lies beneath – might offer some hope for progressive politics. A permanently unstable and mutable location that cannot possibly afford the kind of stability that might be required for proper policing. Yet, it is the very mutability, indeterminacy and fluidity of bodies that seems to provoke policing – and reactions against that policing. We follow this line of argument into the next chapter, which discusses T. E. Lawrence's struggle over skin and identity.

Chapter Three
The Chafing of Bodily Regimes: Skin and the Corporeal Model of the Ego[1]

Introduction

We have heard so far in this book that bodily regimes seek to assign bodies to proper places. Bodily regimes use the bodies of groups and individuals and also body parts to operationalise these assignments and police places. Body parts, the senses, bodies as a whole, things associated with bodies (such as clothes or passports) can all be mobilised to define bodies of groups and of individuals. In the Chapter 2, I argued that bodily regimes are not singular and therefore that bodies can easily be out of place. We also saw that bodies can be indeterminate, meaning that they can cross visible and invisible boundaries that mark proper and improper places. We established that bodily regimes racialise bodies and space. It was suggested that bodily regimes sex bodies and space. Thus, bodily regimes both establish the means of establishing the race and sex of bodies while at the same time working to make race and sex seemingly natural. Indeed, bodily regimes also work through class and sexuality. This chapter seeks to draw out these aspects of bodily regimes more strongly by focusing on the experiences of an upper-class white man, T. E. Lawrence. We follow him as he moves between English and Arab bodily regimes.

[1] Source: From Pile, S. (2011). 'Spatialities of Skin: The Chafing of Skin, Ego and Second Skins in T.E. Lawrence's Seven Pillars of Wisdom', *Body & Society*, 17(4), 57–81. © 2011, SAGE Publications.

However, in keeping with the overall argument of the book, categories of race, sex, class, sexuality are not taken as a given. By tracking T. E. Lawrence in different places, we can illuminate not only the different (English, Arab) bodily regimes as he experienced them, but also his sheer discomfort with the categories of identity that sought to define him. In some ways, Lawrence learns to revel in his discomfort, playing upon his own indeterminacy to find pleasure in poking fun at bodily regimes, and even acting out his discomfort sexually. Significantly, this discomfort reveals itself both consciously and unconsciously in his words and deeds. It is possible to witness this most clearly in his experiences of skin. In this, he appears on the surface to share something with Frantz Fanon.

In Chapter 2, Fanon encountered a white boy on a train in France. In this chapter, T. E. Lawrence encounters British military policemen on a train in Arabia. In some ways, we could see these encounters as a mirror images of one another. A white boy exclaims surprise and fear at the sight of a black man. Indeed, Fanon may have been in uniform, to add to the confusion. A white man challenges a man in Arab clothing for his identity papers. In both situations, the identity of the man is under question. In both, we have bodily regimes that seek to define a proper place for the black man and the man in Arab clothing: neither appear to be in the right place; both are treated with suspicion. But, as we will see, in his situation, Lawrence is able to shift easily back into his assigned place, whereas Fanon cannot find a proper place. What we are tracking, then, is Fanon and Lawrence shifting between bodily regimes. They are both navigating bodily regimes that place them both inside and outside at the same time. Fanon seeks his own bodily regime. Lawrence, however, seems more comfortable being between regimes: he seems to actively seek a location that is itself dislocated, whether it is 'inside' the British military or 'outside' British social conventions of nation, race, gender and sexuality. Lawrence seeks to occupy spaces that enable his suffering: this includes his skin. But there is no proper place for his suffering.

In many ways, we can see attempts (in fantasy) to find solutions to not having a proper place in both Nella Larsen's *Passing* and T. E. Lawrence's *Seven Pillars of Wisdom*. As their characters (and here, I am underscoring Lawrence's own fictionalising of himself) move between worlds – between the bodily regimes that seek to create and define them – we discover them seeking ways to create and define themselves according to new schema. In Larsen, we heard about a colour-rich epidermal schema. In Lawrence's case, we will discover a 'skin for suffering'. Two issues therefore drive this chapter. First, there is Lawrence's movement between worlds, between bodily regimes. Second, there is a more conceptual question about how we understand Lawrence, not as a fully integrated, coherent subject, but as someone who is in parts and between parts that do not quite add up. To achieve this, I discuss Anzieu and Freud's idea of the skin-ego. Significantly, these ideas draw on a spatialisation of the ego and the body.

Skin and the Spatiality of the Ego

In his 2009 paper on the psychoanalysis of skin, Marc Lafrance recovers the work of psychoanalysts Esther Bick and Thomas Ogden for cultural theorists working on the relationship between body, self and society (Lafrance 2009; following Bick 1968; and Ogden 1989). Lafrance observes that skin has become an increasing object of fascination for culture in general (c.f. everything from tattooing to skin beauty products to self-harm to autism), and for cultural theorists in particular (such as Prosser 1998; Benthien 2002; Ahmed and Stacey 2001; and, Connor 2003). Where the work of Bick and Ogden becomes useful, Lafrance argues, is in providing a model of the ego that proceeds from the human experience of skin: that is, in providing a 'psychoanalysis of skin', as he calls it (2009, p. 5; see also Cavanagh, Failler and Hurst 2013; and Diamond 2013).

Lurking in the background of Lafrance's recovery of the 'psychoanalysis of skin' is (as you might expect) Sigmund Freud's notion of the ego, as outlined in 'The Ego and the Id' (1923). We are lucky enough to have two translations of 'The Ego and the Id' (1984 and 2003). It is now possible to work with both these to open up fresh lines of argument in a psychoanalytic understanding of the ego, the body and their surfaces. For Freud, the ego is not only both psychical and bodily, it is also thoroughly spatialised (see Pile 1996). Of particular interest, in this chapter, is the way that the ego is spatialised, through the surface of the body. As a surface entity, the ego enfolds and wraps over as much as it provides cover or creates depth. Such an understanding enables a variably spatialised interpretation of the role of skin, and the idea of a skin-ego, in later psychoanalytic thought. So, one purpose of this chapter is to open up the spatial thinking that necessarily underpins any discussion of the relationship between skin and ego – and, consequently, between bodily regimes and identity formation.

Importantly, psychoanalysis undermines any assumption that skin and ego are necessarily spatially congruent: that is, that the ego is simply bounded by, and contained within, the surface of the skin. Thus, in response to Lafrance's paper, Dee Reynolds concludes that dance can usefully enhance a psychoanalytic understanding of skin and its sensory systems, in part by extending the idea of the body beyond the boundary of skin (2009, p. 30). Dance, Reynolds argues, plays at and with the boundary of skin in ways that require not simply more elaboration, but potentially a new understanding of the dynamic surfaces of skin, ego and second skin. Dance, of course, is not the only place where skin and our understandings of skin rub up against one another.

In this chapter, I have chosen Lawrence of Arabia's war-time experiences, as told in his autobiographical *Seven Pillars of Wisdom* (1926). Here, I am seeking to do a little more than add another example to the list. I wish to use Lawrence's experiences *both* to enrich the ways that we understand skin, ego and second skins, *and also* to better see how these might chafe against one another. Anzieu's notion of a skin-ego is particularly helpful here (1985; 1990): not only does he show that the surface spaces of skin, ego and second skin can be fluid, porous and colander-like, Anzieu also points to those moments where skin-egos are 'toxic'. Of particular interest, here, is Anzieu's discussion of Monsieur M, whose chafing of skin and ego is especially severe. This chafing is also evident in Lawrence's case, not the least because his life is also highly territorialised – and not simply as a direct conse-quences of fighting for the British with an Arab army during the First World War.

Thinking through skin, it is argued, means working at the surface of the body, a dense surface where the social, the psychological and the bodily are inseparable (see Gallop 1988; and also Ahmed and Stacey 2001). Thinking through the ego, as we will see, similarly means working at the surface of the body, also where the social, the psychological and the bodily are inseparable. Intriguingly, then, both skin and ego are dense surfaces, affording a contact zone between body and society. To understand this, I outline the notion of the ego as a psychological manifestation of body and of skin, drawing upon Sigmund Freud, Esther Bick, Thomas Ogden and Didier Anzieu. While the psychoanalytic model of skin, ego and second skins is fundamentally spatial, the lived experiences of those spaces – that is, their spatialities – are variable and dynamic (see Pile 1996, 2009). So, this chapter deploys a spatial vocabulary based not simply upon terms such as space, territory, ground, location, surface, boundary, inside/outside and the like, but also on notions of fluidity, wrappings, wraps, porosity, holes, rips, thickness/thinness, patching – and especially chafing – and the like.

To develop an understanding of the lived spaces of skin, ego and second skins, I discuss Lawrence's experiences in Arabia. Significant, for my purposes, are two specific incidents that Lawrence describes in *Seven Pillars of Wisdom* (1926): the first occurs when Lawrence's identity is questioned by British military police; the second, when Lawrence is tortured by the Turkish army in Deraa. In each of these incidents, I argue, it is Lawrence's skin that is at stake. It is through Lawrence's description of these incidents that the spatialities of skin, ego and second skin can be elaborated.

The Spatialities of Ego, Skin and Second Skin

As Ulnik (2007) shows, psychoanalysis has engaged extensively with the problem that skin set for psychological and bodily development. Thus, for Freud, skin is an erotogenic zone; a source of, and object for, a drive to touch; a site of contact and contagion; a boundary layer representing protection and vulnerability; a source of sensations; a dense site where physical dis-eases intertwine with psy-chical symptoms, and more (Ulnik 2007, ch. 1). Nonetheless, it is with Anzieu (1985) that skin is given especial prominence in the development of the ego. Of particular interest is the way in which both Freud and Anzieu spatialise the ego through an understanding of it as a surface, or skinly, entity. Here, Freud's paper on 'The Ego and the Id' is absolutely fundamental (1923).

Freud asserts:

> The ego is first and foremost a bodily ego; it is not merely a surface entity, but is itself the projection of a surface. (1923/1984, p. 364)

In a footnote for the 1927 English translation, Freud elaborates:

The ego is ultimately derived from bodily sensations, chiefly from those springing from the surface of the body. It may thus be regarded as a mental projection of the surface of the body. (1923/1984, pp. 364–365)

As if to reinforce the point, on page 366, Freud asserts that the conscious ego 'is first and foremost a body-ego'. There are three conclusions to be drawn from Freud's understanding of the ego: first and foremost, it is bodily; second, it is a surface entity; and, third, it is a mental projection of a surface. The ego, then, is a body-surface-projection. Since the ego is a mental projection of the surface of the body, it is worth asking which surface. It seems unlikely that Freud was referring either to specific sensations (such as warmth or comfort or taste) or to particular parts of the surface of the body (such as the face or mouth or genitals). Rather, Freud is making a connection between, on the one hand, the range of sensations that skin provides, and, on the other, the fact of skin as a surface covering the whole body. The ego, in this light, is skinly – a mental projection of skin. As a mental projection, the ego takes on the shape and function of skin. Thus, the ego is ultimately a psychic skin derived from bodily sensations mediated, primarily, through skin itself.

At first glance, it may appear that the ego provides a psychic skin that, like skin itself, both covers and coheres the body and also provides the developing child with a secure ground for its growing sense of self. However, Esther Bick's meticulous observations of infant development and the care-giving environment contradict this cosy reading of the skinly development of the child (Bick 1968; see also Rustin, 2009). Bick understood the infant as occupying two contradictory positions. On the one hand, feelings of being-in-the-world derived from the body, and the liveliness and integration of the body, lend the infant a sense of its own *coherence*. On the other hand, the baby experiences periods of anxiety and disintegration, which give the infant a sense of its own *incoherence*. For Bick, the presence or absence of the (main) caregiver is pivotal for the infant's oscillation between feelings of coherence and incoherence.

Of particular interest to Bick was how the baby would create a sense of coherence for itself after periods of incoherence, how it would bind itself together. Famously, Bick argues:

[. . .] in its most primitive form the parts of the personality are felt to have no binding force amongst themselves and must therefore be held together in a way that is experienced by them passively, by the skin functioning as a boundary. (1968, p. 55)

Thus, the skin is a boundary projected inwards by the infant to bind together, to enfold and contain, the myriad parts of its bodily experiences, in the face of its commensurate experiences of anxiety, loss and disintegration. Bodily sensations from skin provide the infant with a systematic framework *both* for containing and protecting its developing sense of itself, *and also* for developing an ego and a sense of self. Moreover, skin provides the infant with a sense of integration, continuity, boundedness and containment, such that it knows where it is (albeit inevitably associated with anxieties over not being anywhere in particular). Thus, the where of the body (and especially its surface) provides the ground on which the baby can become who it is. Significantly, and seemingly following Freud, Bick argues that this ground is a product of fantasies as much as bodily sensations. The infant produces for itself a world modelled on the two sides of skin: one, inward facing; the other, outward facing.

For Bick, the world is divided into interior spaces and exterior spaces (1968, p. 56). This 'split' enables further differentiation of interior and exterior worlds through psychic splitting and idealisation (following Klein). Significantly, Bick argues, the baby uses the model of skin to patch up parts of itself – either where there is a failure to contain its parts or where there are gaps or weaknesses (or thinnesses) in its bodily or psychic skin. These patches Bick calls 'second skins' (pp. 57 and 59), and these second skins can take an endless variety of forms (see also Lafrance 2009, pp. 9–10; Kogan 1988; and Prosser 1998).

Similarly to Bick, Ogden (1989) sees skin as providing the infant with a sense of ground (Lafrance 2009, p. 12). What Ogden adds to the analysis of the relationship between skin and self is a deeper understanding of primitive sensory experiences. For Ogden, the child's world is initially shaped by its senses and not by emotion or affect, such as love or envy (as much psychoanalysis would argue). Unable to distinguish between interior and exterior worlds, the baby builds its sense of self out of its experience of 'shape and rhythm, pattern and periodicity, texture and temperature' (Lafrance 2009, p. 13) – particularly as these are experienced through the skin. Thus, the surface of skin sensations provides a ground upon which the infant fashions its sense of self; upon which the infant anchors and organises its experiences of the world; and, upon which it differentiates between interior and exterior worlds.

Like Bick, Ogden also suggests that 'second skins' are formed to provide 'epidermal armouring' as a defence mechanism against anxieties, but also a means to prevent dissolution, shapelessness and loss. Second skins, as for Bick, become a way to bind together their sense of self along lines offered by the infant's experiences of skin. Even so, these second skins take many forms, drawing upon senses of smell, heat, touch, hearing and so on. What these second skins reveal are:

> the lengths to which human beings will go in order to feel enclosed and enveloped by their bodily surfaces. (Lafrance 2009, p. 17)

This insight will help us understand some of Lawrence's experiences in Arabia. In many ways, it is possible to argue that Lawrence, in order to feel enclosed and enveloped, creates for himself a particular form of skin wrapping – a *skin of suffering*, to echo Anzieu's interpretation of skin functioning as a 'wrapping of suffering' (1985, ch. 16). This skin of suffering provides the ground upon which he fashions his military body, his militarised ego. Lawrence can be seen as someone who seeks to establish and re-establish his sense of self through privations meted out on the surface of the body: that is, on his skin. Yet, there is also a geopolitics of skin here, emerging from the conjunction of a European War spreading beyond Europe; British, French and Turkish colonial ambitions in Arabia; and, Western Orientalism. Specifically, Lawrence's white skin differentiates him from Arab skin; yet, his adoption of an Arab second skin also marks him out from the British. Moreover, Lawrence's skin asymmetrically – British-Arab/ Arab-British – positions him in relationship to his sense of self. Nevertheless, the idea of a skin of suffering might still help us understand why Lawrence is always on the verge of disintegration and dissolution; why he is unable to provide himself with a secure enough ground that would enable him to adopt a secure place in the world, personally, politically, culturally, corporeally.

To understand Lawrence, however, we need to revisit the surface of the body in the conception of the ego. It is not simply the case Lawrence's skinly ego grounds or envelops him; nor even that his skin is a boundary or wrap for his body. Instead, Lawrence appears to have many surfaces, none of which quite cover him. To get at these aspects of the ego, it is worth returning to Freud's definition of the ego as a body-surface-projection. He says (alternative translation):

> The ego is above all a *corporeal* entity; it is not merely a surface entity, but is itself a *projection* of a surface. (1923/2003, p. 117: note italics in this translation)

Commenting on this, and the famous 1927 footnote, the translator John Reddick is at pains to point out that 'projection' has specific meaning for Freud (p. 258): in early twentieth century neurophysiology, *projection* refers both to the spatial distribution in the brain where neural impulses are registered, and also to the network of fibres through which impulses pass to the brain. Projection, in this view, is not merely fantasised, as if projecting an image onto a two-dimensional screen, but the product of the neural 'hard-wiring' of a three-dimensional body – as in Penfield's homunculus, which sculpts the human body to reveal the physicality of its sensory networks, such that the 'little man' has disproportionately large lips, hands, feet and genitals.

So, in this understanding, the anatomical location of consciousness and of ego – which is modelled on skin – is not simply *on* the surface, but a surface produced by the hard-wiring of the body, which has properties (at least) of volume, shape and surface. It also provides a sense of location in the world as well as an orientation to that world, where orientation is itself multiple, including directionality (towards/away from), proprioception, a sense of movement, relative space (front/back, near/far, etc.), location (here/there) and so on. In this understanding, the ego should be thought of as being like a body – literally body-shaped – and not merely being of the body or simply of a surface. Similarly, Freud argues that

> consciousness constitutes the outer surface of the psychic apparatus, which is to say that we have defined it as a function of a system that is spatially the closest to the external world – as spatial proximity, incidentally, that applies not only in terms of function but also [. . .] in terms of anatomical location. (1923/2003, p. 110)

The body is the model for the mind: the surfaces of the mind modelled on the surfaces of the body. This maps directly onto Anzieu's notion of 'skin-ego' (1985), so that the ego is modelled on the child's experience of skin and is also a mental projection of skin. For Anzieu, moreover, the ego is structured like and functions as a skin (1985, p. 44; also see pp. 14–16); having functions such as holding, containment, protection, individuation, 'common sense', sexual excitation, the distribution and organisation of libido, an awareness of reality and social being, and (importantly) self-destruction (see Ulnik 2007, p. 64). Thus, exactly like skin, the ego acts as a boundary, a wrapping, an interface, a protective covering, a permeable layer, a filter, a sense organ, and so on. Directly echoing Freud's definition of the body ego, Anzieu defines *skin-ego* as:

> a *mental image* used by the child's Ego during its early stages of development to *represent itself* as an Ego containing psychical contents, based on its experience of the surface of the body. (1985, p. 4, emphasis added)

Similarly, elsewhere, Anzieu argues, paraphrasing Freud, that 'the unconscious is the body' and thus that, contra Lacan, the unconscious is structured like the body (1990, p. 43). The body, in this conception, is not simply anatomical, it is also the product of phantasy and infantile sexual theories and a source of experiences, perceptions and thoughts. The advance Anzieu makes on Freud is not, simply, an expanded list of functions for skin-as-ego and ego-as-skin. Like Freud and Bick, Anzieu stresses that the skin-ego is an interface, boundary, frontier or layer between the inside and outside, between the individual and the world. However, his notion of wrapping allows for this layer – and the surface of the skin – to take markedly different shapes, drawing on skins many sensibilities (touch, sound, pleasure/pain, taste, heat, muscularity). This provides a direct link with Rancière's distribution of the sensible (2004), yet reveals how skin actually generates many schemas for organising the sensible and for generating sensibilities. Moreover, whilst the stress on surfaces might imply the model of the skin-ego is two-dimensional, Anzieu emphasises the three-dimensionality of psychical and bodily space, even as these emerge as surfaces from surfaces (Anzieu 1985, p. 263). This is clear in the case of Monsieur M (Anzieu 1985, pp. 118–120).

Monsieur M is a radio electrician. He is also a masochist. For Anzieu, Monsieur M's masochism is an attempt to re-establish the functions of the skin-ego, and he uses his skin as a means to achieve this. Monsieur M's methods are tenacious. He has, for example, inserted bits of metal and glass under his skin, including needles in his penis and testicles; 'by cutting thongs out of the skin of his back [. . .] he could be suspended from butcher's hooks while being sodomised by a sadist' (Anzieu 1985, p. 118). His body shows innumerable wounds and scars: Monsieur M has, for example, torn off his right breast and sawn off the little toe on his right foot. For Anzieu:

> The containing function [of the skin-ego] was restored by the repeated creation of a wrapping of suffering, through the great variety, ingeniousness and brutality of instruments and techniques of torture: a perverse masochist has to keep the fantasy of ripped skin perpetually alive in order to re-appropriate the Skin-ego. (1985, p. 118)

How we understand the idea of the *wrapping* in Anzieu's work becomes key to understanding how a three-dimensional homunculus comes about from topologically two-dimensional projections and surfaces. Throughout Anzieu's analysis of skin-ego, he discovers a variety of wrappings through which people attempt to establish and re-establish their skin-ego, as a surface of body-and-mind. For example, there are narcissistic, psychical, hysterical, excitation, anxiety, suffering and masochistic wrappings. The term *wrapping* has both two-dimensional and three-dimensional properties. On the one hand, it is simply a surface; on the other hand, as it wraps and enfolds, the surface provides shape and volume. Significantly, these wraps are not necessarily complete, nor are they indestructible, nor are they simply singular – even while one takes precedence over all others, as in the case of Monsieur M.

From this perspective, we can suggest that the mental image of skin-ego is a three-dimensional corporeal model built out of the infant's experiences of its bodily sensations and surfaces. A modified definition of the skin-ego, following this line of argument, would be this. The skin-ego is the sensorial homunculus that the child uses to present and represent itself, and upon which it anchors and organises its experiences of the world. More than a container or boundary layer or frontier even, the skin-ego is lumpy, misshapen, multiple and unevenly developed – and from the very beginning. There are, moreover, many surfaces in the body that can provide models for the ego. For example, Anzieu mentions the experience of sensations drawn from muscle (1985, ch. 15) and from sound (1985, ch. 11), but there are more, such as smell and taste, but also breathing, feeding, digestion, and so on. Moreover, skin sensibilities have a variety of different relationships to the 'corporeal model' ego. Not merely binding together the skin-ego, nor providing an armour that protects the surface of the ego, the sensible skin might effectively become an alternative homunculus – it might, indeed, grate against the skin. Indeed, as we will see, skin, skin-ego and second skins can cause friction with one another: in other words, following Lawrence of Arabia's experiences, these different corporeal models of the ego can chafe against one another.

This discussion of a bodily model for the ego has implications for how we understand bodily regimes. To begin with, we can see that bodily regimes are linked to people's psychodynamics, involving both how people come to understand themselves and how people come to live their bodies. Both, consciously and unconsciously. The aim in this chapter is to think through the relationship between the skin-ego (identity formed through the skin), but we are clearly building a model for the ego where neither body nor identity is singular. This multiplicity, most importantly, is not simply a list of identities associated with aspects of the body-ego, for these relationships are underpinned by unconscious dynamics. The lesson here is that bodily regimes operate through conscious and unconscious processes of identity formation. Significantly, this discussion of the unconscious dynamics of skin allows us to see how corporeal schemas and bodily regimes might be generated in ways that might run counter to common sense and not align in a strong way with the 'official' distribution of the sensible (as Rancière would have it), whatever that might be. Indeed, it is not possible to understand Lawrence of Arabia's experiences if he is simply seen as a location within a flat distribution of the sensible.

In Freud and Anzieu, skin is a surface that takes on volume, shape and size: modelled on the body, for sure, but not on the physical body as a superficial, two-dimensional entity. To this, we can usefully add Ogden's senses of the body. Taken together, psychic space and bodily space ground the skin-ego in many spatialities at once, such as: surface and depth, volume and texture, temperature and thickness, rhythm and proprioception, wraps and gaps, ease and disease, and so on. Where skin is understood as a surface, second skins are often seen as re-establishing or repairing the surface of the skin-ego. However, if the skin provides volume and shape, without necessarily containing or bounding the body – and skin, after all, is experienced as having many openings: on the face alone, nose, eyes, mouth, ears, tear ducts and pores – then the roles of second skin would include the provision of volume and shape to the ego as well as porosities and openings out to the world.

I have suggested that a productive way to understand Lawrence's experiences in Arabia is through the idea of a skin of suffering. It is possible to argue that Lawrence's skin of suffering is a psychical defence mechanism designed to re-establish his bodily ego (following Anzieu 1985, ch. 16). Suffering would, in this view, provide Lawrence with a sense of continuity or cohesiveness or boundedness; of his interiority; and, a clear distinction between internal and external worlds (Anzieu 1985, p. 160; similarly on p. 136). To understand this, Anzieu uses an analogy with the Möbius strip, which I discuss in depth in Chapter 4. From this perspective, Lawrence seeks suffering because it secures and re-secures the ground upon which he has fashioned his skin-ego (see also Benthien 2002, p. 120).

However, Lawrence's skin of suffering does not appear to enfold or contain him, nor repair the holes and fissures in his skin-ego. Instead, he appears to actively seek his own dissolution, disintegration and decoherence. On the other hand, his second skins appear to reach out to others, whilst also reaffirming his own disaffection and discomfort. Skin and second skin appear to chafe against one another. To understand this, I now turn to Lawrence of Arabia's experiences of skin, ego and second skin. First, a little background.

Lawrence of Arabia

Lawrence of Arabia remains a name to conjure with. The story of the British intelligence officer who lived among the Bedouin Arabs, became a commander of their guerrilla army, and led them to freedom from Ottoman tyranny during the latter part of the First World War, has proved to be one of the enduring myths of military manhood in twentieth-century Western culture. (Dawson 1994, p. 167)

And, we should add, twenty-first century Western culture: For example, there is Rory Stewart's two-part television series, *The Legacy of Lawrence of Arabia*, broadcast on BBC2 on 16 and 23 January 2010.

Thomas Edward Lawrence (16 August 1888 to 19 May 1935) – also known as T. E. Shaw (a name adopted in 1923, after an attempt to pass under the name John Hume Ross failed) – is best known as *Lawrence of Arabia* (for full biographies, see Wilson 1989; or James 1995). Like many, I am particularly interested in the Deraa incident, but I will also explore Lawrence's return to British Headquarters in Cairo after the capture of Akaba. (I use Lawrence's spelling of place names: where appropriate, contemporary spellings are provided in square brackets.) In both episodes, Lawrence's skin plays a significant, yet overlooked, part in the story. In Deraa, it will add to our understanding of Lawrence's sexuality and masculinity; in Cairo, it will show us how Lawrence's body refused him a proper place and how he used his body to refute his assigned proper place. Though skin can be mutable (as we have seen), Lawrence reinforces the point that skin is far from being simply mutable or, put another way, far from simply being a surface upon which meanings, culture and value are socially inscribed.

Lawrence was born out of wedlock in 1888. He was the son of an aristocratic Anglo-Irish father, Sir Thomas Chapman, and his daughter's former nursemaid. Educated at Eton, Lawrence was subsequently awarded a First Class degree in History at Oxford. His interests took him to the Middle East in 1909, where he began to research a doctoral thesis on the architecture of Crusader Castles. Between 1910 and 1914, he worked on an archaeological dig at Carchemish, under David George Hogarth and R. Campbell Thompson. In 1914, with the outbreak of the First World War, Lawrence went to work with British intelligence at Cairo's map department.

In June 1916, an Arab Revolt against the Turkish Ottoman Empire began when Mecca was seized by Grand Sharif Hussein ibn Ali, ruler of the Hejaz Arabs. The Bedouins immediately sought help from the British in their fight against the Turks. For the British, locked in the brutal stalemate of trench warfare on the Western front, this was an opportunity to undermine Turkish support for the Germans. In particular, the British were interested in disrupting the railway supply lines connecting Arabia to Turkey. For once, it seemed that British and Arab interests coincided. In October 1916, Sir Ronald Storrs was sent to the Hejaz region to assess the situation. He was accompanied by Captain T. E. Lawrence.

Of course, what happened in the ensuing two years is the stuff of legend (see Thomas 1924; or Korda 2010). Lawrence's respect and love towards both Arabs and Arab culture is part and parcel of this legend. For some, this love and respect was an outcome of Lawrence's location within Orientalist discourse, which exoticises and eroticises the Arab while remaining flexibly superior to it (see Said 1978, pp. 228–231; or Deleuze 1998). For others, it is a product of the paradoxes of class, masculinity, race, gender and power in a colonial setting (see, e.g., Silverman 1992, pp. 337–338). Seeing Lawrence these ways tends to convert him into a sign of his time and place. Lawrence himself did not seem particularly comfortable with either his time or his place.

Indeed, Lawrence sought to 'correct' the popular image of him as the heroic military genius, by providing his own account of himself: the autobiographical *Seven Pillars of Wisdom* (initially privately published in 1926). In this book, Lawrence provides an account of his times and his geographies, laying great emphasis on the description of the desert, of his relationships to the people around him, and of his understanding of the nature of warfare (see also Johnson 2020). Arguably, the narrative is as much legend as history. But, whatever one might say about the truth, self-abasement and self-aggrandisement in his account, it is impossible to reduce Lawrence to one overriding narrative: even his own is riddled with contradiction and confusion.

Historian, warrior, spy, Imperialist, freedom fighter, fantasist, victim: Lawrence ultimately refuses to settle easily into (post)colonial grids of class, sexuality, race, gender and power. Put another way, Lawrence seemed to occupy many spaces and points in the maps of identity and power. These do not cohere or consolidate around one particular space (e.g. British military space) or location (e.g. Imperial hero). An attendance to Lawrence's lived experiences of his spaces and skins suggests more than a refusal to settle into one space and one skin, nor one bodily regime and one proper place, however. Rather, we can glimpse not only how Lawrence revelled in his awkward adoption of different spaces and skins, but also how these spaces and skins chafed against one another.

'What Army, Sir'?: A Question of Second Skins

By the end of November 1916, Lawrence had adopted full Arab dress at the request of Prince Feisal – third eldest son of Sharif Hussein bin Ali. Feisal is a significant figure: after the war, Feisal was briefly King of Syria before becoming King of Iraq. Feisal insisted Lawrence wear Sharifian robes, with a gold dagger, that would distinguish him as a man of wealth and influence. Although Lawrence, in fact, had little choice, he gladly accepted, partly because he thought it would help put the Arabs at ease. It was an anti-Imperial gesture, for the only other people in khaki uniform that the Arabs had seen were Turkish Officers of the Ottoman Empire (Lawrence 1926, p. 129). Lawrence's Sharifian robes were more than simply a costume, a way to fit in, a political or military expedience, or 'drag'; he had a deep fascination for Arab cultures and ways of life and it is easy to see his adoption of Arab dress as the adoption of a new second skin for his body and his sense of self. For him, being amongst Arabs conjured up images of the crusades, fierce and brave warriors, an ancient civilisation steeped in nobility and honour. Unsurprisingly, he felt proud to wear Arab clothing.

In fact, it was common for British liaison officers to wear Arab dress, as did Lawrence's own superior officer Colonel Stewart Newcombe. What set Lawrence apart was his seeming whole-hearted embracing of an Arab way of life, and also his passion for the cause of Arab liberation. Nevertheless, Lawrence was all too acutely aware of the tensions of identity that this created in him, and for others:

> In my case, the efforts for these years to live in the dress of Arabs, and to imitate their mental foundation, quitted me of my English self, and let me look at the West and its conventions with new eyes: they destroyed it all for me. At the same time I could not sincerely take on the Arab skin: it was an affectation only [. . .] I had dropped one form and not taken on the other [. . .] Sometimes these selves would converse in the void; and then madness was very near, as I believe it would be near the man who could see things through the veils at once of two customs, two educations, two environments. (Lawrence 1926, p. 30)

The affectation of Arab culture left Lawrence feeling neither Arab nor English. Though Lawrence evidently experienced himself in the void between worlds, Arab and English, he moved in both. Yet, adding to his feelings of being in a void between worlds, Lawrence knew he was only partially accepted in each. Quitted of his English self, and unable to experience his Arab second skin as anything more than an affectation, Lawrence was left feeling close to madness. Put another way, the psychological and physical apparatus of British/Arab second skins were failing to provide Lawrence with a singular sense of self. Instead, the wraps of British and Arab skin and second skin only partially cover him.

Worse, by (repeatedly) switching from one second skin to another, Lawrence had de-realised both – in his eyes, everything was destroyed. Yet, Lawrence ultimately chooses the void, the edge of madness – for Lawrence neither 'goes native' at this point, nor retreats back into the safe ground of British militarism. Strangely, then, what Lawrence seems to adopt is *the failure of his skins* – more than this, even, he revels in their failure. We can see this in Lawrence's description of his arrival in Cairo at the British headquarters.

Perhaps Lawrence's most famous, and remarkable, military exploit is the taking of the strategically significant port town of Akaba [Aqaba] on 6 July 1917. Lawrence rushed to inform General Allenby that Akaba had fallen. He had been on active duty for four months: continually on the move, suffering constant privation. Lawrence writes:

> In the last four weeks I had ridden fourteen hundred miles by camel, not sparing myself anything to advance the war; but I refused to spend a single superfluous night with my familiar vermin. I wanted a bath, and something with ice in it to drink: to change these clothes, all sticking to my saddle sores in filthiness: to eat something more tractable than green date and camel sinew. (1926, p. 326)

On the way to Cairo, Lawrence's sufferings were eased with the aid of an intelligence officer, Lyttleton. Lyttleton quickly set up Lawrence with a room at the Sinai Hotel. Initially, staff there reacted with hostility towards Lawrence and his filthy (Arab) appearance. Lyttleton was even informed by his own spies that 'a disguised European' was in the hotel. Eventually, however, Lawrence got the hot bath and ice cold drink that he craved.

The next day, Lawrence set out for Cairo by train. Although provided with tickets and passes, he was stopped by British military police. Mistaking him for an Arab, they wanted to know who he was.

> The strenuous 'control' of civilian movement in the canal zone entertained a dull journey. A mixed body of Egyptian and British military police came around the train, interrogating us and scrutinising our passes. It was proper to make war on permit-men, so I replied crisply in fluent English, 'Sherif of Mecca – Staff', to their Arabic inquiries. They were astonished. The sergeant begged my pardon: he had not expected to hear. I repeated that I was in the Staff uniform of the Sharif of Mecca. They looked at my bare feet, white silk robes and gold head-rope and dagger. Impossible! 'What army, sir?' 'Meccan.' 'Never heard of it: don't know the uniform.' 'Would you recognise a Montenegrin dragoon?' (1926, p. 327)

Lawrence is evidently very pleased with his ability to fool the military policemen, by playing the void between identities, Arab and English. His voice English, with a shrill Etonian accent; his dress impressively Arab: his joke assured. He is laughing at the 'permit-men'. He is also laughing at the possibility that they could possibly know, for definite, who they were talking to. Yet, under the scrutiny of white eyes, Lawrence turns Arab, not because of his raw skin or his matted blonde hair, but because of the second skin of his dress: white robes effectively disguise the white skin of his body, making him Arab.

Lawrence's laughter contrasts with Fanon's response to being identified as a Negro (also, curiously, on a train whilst on military duty): Fanon tries to laugh, but cannot (Fanon 1952, p. 111; see Chapter 2); while Lawrence becomes his second skin, Fanon's is irrelevant. In *Black Skin, White Masks* (1952), Fanon argues that black skin becomes the outstanding sight (and site) of difference and inferiority in a white-dominated world. This is to say that colonial power operates through a corporeal schema, which grades bodies according to the colour of the skin. It should be noted that blood, hair, bones and so on, are also chained (through the body) to skin in these racist bodily schemas. For Fanon, the colonial regime's imposition of skin hierarchies not only defines the visibility of the body, but also territorialises the body, giving bodies proper, or improper, places to be (p. 111; see also Pile 2011). Thus, Lawrence's laughter enables him to occupy his proper place, to surprise and thwart the permit men's imposition of a British/Arab bodily schema, while Fanon's initial inability to laugh reveals that he is not in his proper place – and that is no joke.

For Lawrence, colonial, military ways of checking identity by sight are, simply, absurd. He knows he looks like the wrong person in the wrong dress in the wrong place. He also knows that identity cannot be read through corporeal schemas that rely on solely visual cues, whether these are found in clothes, skin or hair. Lawrence rails against the 'visual skin' of colonial expectations by using his incongruously English voice. Thus, Lawrence knowingly uses his out-of-place voice to challenge the distribution of the sensible that the military policemen both inhabit and literally police. In his laugh, Lawrence's 'skin of voice' chafes both against the visual skin of racist colonial schemas, which reads people through their visible surfaces, and also against the surface of his own second skin.

Lawrence plays on the deep uncertainty that the Arab body represents for the British military police (compare Bhabha 1994, p. 44) – albeit using the privilege of his race. Yet, Lawrence's laughter might also be thought of as slightly mad, poking fun at his own side, his own people; treacherously, and unrepentantly, undermining the authority of British soldiers in front of their Egyptian counterparts. Lawrence wanted to be the sore thumb that stood out. On arrival at British military headquarters, he is notably dark, short and scruffy – in Arab clothing. He does not show his face. Even after they had seen beyond the Arab dress, senior officers remained disconcerted by Lawrence's crimson skin and haggard looks (1926, p. 328).

Lawrence seems determined to make his skin and second skins as uncomfortable as possible. Rather than wrapping him up, and, thereby, however perversely, re-establishing his skin-ego, Lawrence's skins and second skins are at odds with one another. It is not simply that his skins do not fit, they also provide a means of suffering; not only is skin a location for this suffering, so are the contact points between his skins and others.

Skin of Suffering: The Deraa Incident and Lawrence's White and Fresh Skin

The Deraa [Daraa] incident is one of the most controversial episodes in Lawrence's wartime experiences, not the least because Lawrence continually revised the story (see James 1995, pp. 245–263). It is, indeed, most likely that the incident did not take place (see Barr 2006), or at least that it did not take place there and then. In *Seven Pillars of Wisdom* (pp. 450–456), Lawrence recalls that, on the night of 21–22 November 1917, he was captured, interrogated, whipped and raped by the Turks in Deraa. Though badly injured, Lawrence managed to escape to Akaba. Some commentators have observed that Lawrence, in fact, displays a homoerotic fascination with Arab homosociality and Arab male beauty (see James 1995, p. 253; see also Silverman 1992, pp. 299–338). Typically, such authors have seen Lawrence's account as a masochist fantasy, as prefiguring his actual (later) masochist sexuality and/or as evidence of his homosexuality. At stake in these debates is (a) the question of whether this is an Orientalist fantasy, or (b) Lawrence's identity as a solider hero, untarnished by sexual innuendo and perversion, or (c) both. For Deleuze, meanwhile, Lawrence's shame (at being raped) coexists and intermingles with his sense of glory (1998, p. 124), which can be seen as somewhat typical of imperial (male, military) subjectivity. Building on these readings, let us see what Lawrence's skins tell us.

Accompanied by a boy, Halim, and an old man, Faris, Lawrence was conducting (by his own account) a daring spying mission in the small town of Deraa, looking for military weak points for a planned attack, when they were stopped by a Turkish sergeant. The three – Lawrence, Halim and Faris – were wet from rain and covered in mud; Lawrence himself was struggling with a broken foot. The sergeant accosted Lawrence roughly. He told them the Bey, the local military commander, wanted to see them. Feeling that flight was impossible, Lawrence agreed. At the barracks, Lawrence recounts, a 'fleshy Turkish officer' asks him his name: 'I told him Ahmed ibn Bagr, a Circassian from Kuneitra' (p. 451). The officer did not believe Lawrence and he was taken to the guard room. Soon, 'three dark men' came for Lawrence and took him to the Bey's bedroom. The Bey:

> was another bulky man, a Circassian himself, perhaps, and sat on the bed in a nightgown, trembling and sweating as though with fever. When I was pushed in he kept his head down, and waved the guard out. In a breathless voice he told me to sit on the floor [. . .] At last he looked me over, and told me to stand up: then to turn round. I obeyed; he flung himself back on the bed, and dragged me down with him in his arms. When I saw what he wanted I twisted round and up again, glad to find myself equal to him, at any rate in wrestling. He began to fawn on me, saying how white and fresh I was, how fine my hands and feet, and how he would let me off drills and duties, make me his orderly, even pay my wages, if I would love him. (pp. 451–452)

Lawrence refuses. The Bey orders him to undress, but Lawrence hesitates. Now angry, the Bey calls in a sentry, who tears off Lawrence's clothes 'bit by bit' (p. 452). The Bey now notices the 'half-healed places where the bullets had flicked through [Lawrence's] skin' (p. 452). The Bey then begins to paw Lawrence; Lawrence violently retaliates. Four more soldiers are called, and they take hold of Lawrence by his hands and feet. Using his slipper, the Bey repeatedly slaps Lawrence in the face (a grave insult), while a corporal holds his head so that he cannot flinch. The Bey then 'leaned forward, fixed his teeth in my neck and bit till the blood came. Then he kissed me' (p. 452). The Bey then takes a dagger and begins to push the tip of the blade through folds in Lawrence's skin. Blood flows down Lawrence's side, onto his thigh. The Bey 'looked pleased and dabbed it over my stomach with his finger-tips' (p. 452).

Lawrence, at this point, tells the Bey who he really is (p. 453). To Lawrence's astonishment, the Bey replies that he already knows. The corporal is ordered to take Lawrence out and teach him 'everything' (p. 453). Lawrence is then severely beaten by the soldiers. The soldiers taunt Lawrence that he will not be able to resist a whipping – by the twentieth lash, he would 'beg for the caresses of the Bey'. They then use a Circassian whip on him, each 'madly across and across with all his might' (p. 453). During the beating, the soldiers would 'squabble for the next turn, ease themselves, and play unspeakably with me'. Repeatedly. Until the soldiers felt that Lawrence was 'broken', his skin now covered in welts and blood. He was kicked to his feet by the corporal. Lawrence then recalls:

> smiling idly at him, for a delicious warmth, probably sexual, was swelling through me: and then that he flung up his arm and hacked with the full length of his whip into my groin. (p. 453)

By now, Lawrence's skin was so battered, bruised and blooded that the Bey rejects him. The soldiers complain that the fault lay not with the severity of their beating, but with Lawrence's 'indoor skin' (p. 455). The corporal – 'the youngest and best-looking' of the soldiers (p. 455) – then took Lawrence into the street and released him. Unaccountably, Lawrence is then taken care of by the townsfolk, who wash, drug, bandage and clothe him, such that he can make his escape. But Lawrence remains haunted by these events – 'that night the citadel of my integrity had been irrevocably lost' (p. 456).

It is easy to see how the Deraa incident, as Lawrence relates it, can be understood as a homoerotic masochistic fantasy. Lawrence himself wonders whether it was actually a dream. Indeed, its unreal quality nags at him. To assure himself (and the reader) that he did not in fact fantasise the whole incident, Lawrence repeatedly writes about his bruises and wounds: his skin becomes the irreproachable witness to the truth of his experience. The pain from the bruises and wounds provides a sure ground upon which Lawrence can feel the reality of his experience, however dream-like it seems. The pain disperses his doubts. This makes perfect sense, of course. Yet, Lawrence also confesses to the pleasure of his pain, especially the pain experienced on the surface of his skin.

The whipping, Lawrence confides, aroused him sexually, the delicious warmth hinting at orgasm. This tallies closely with revelations of Lawrence's sexual activities, long after the war (see Simpson and Knightly 1969). Whilst serving with the RAF at Bridlington, Lawrence paid his friend (from their time together in the Tank Corps) John Bruce, for over a decade, to whip him with a birch until

Lawrence either orgasmed or passed out. Bruce himself thought this was an act of shame by Lawrence, but it is more likely this was a replaying, or reliving or re-fantasising of the masochistic eroticism of the Deraa incident. Far from Lawrence's skin providing an uncertain zone of identity, race and sexuality, his skin defines it. Standing naked, sexually aroused, Lawrence was unable to be anything other than himself. And he suffered for it, much to his own pleasure (as Silverman observes, 1992, p. 330).

There is another possible interpretation of Lawrence's supposed new-found masochism. It is worth noting that Lawrence's descriptions of his suffering play out on the skin – the Bey's knife, the beatings, the whippings, the sensation of warm blood and the bruises. His 'white and fresh' skin would have looked, to any Turkish soldier who saw Lawrence naked, completely unlike any Circassian they had seen. His body would have had a brick-red face and hands, yet be an uncooked white everywhere else. Unquestionably, Lawrence's skin would have immediately given him away as an English, or European, spy. In wartime, it is impossible to imagine that Lawrence would have been simply released. This implies that the Deraa incident, *if true*, probably took place before the war, somewhere other than Deraa, perhaps while Lawrence was a student. Arguably, Lawrence was already in a place of suffering long before the war ever started. The war was opportune: it afforded Lawrence both many ways to ensure that his first skin endured enough suffering, and also the singular opportunity to replay the sado-masochistic scene as an unseemly circumstance of war. More than this, however, Lawrence could present Deraa as an antiheroic episode, in which his citadel was lost, thereby reliving the suffering again and again. Through Deraa, Lawrence *doubled*, and *re-doubled*, a skin of suffering for his ego.

There seems, in this, to be an uncanny parallel between Lawrence's 'skin of suffering' and Monsieur M's masochism. Indeed, Monsieur M would also achieve orgasm through having the whole of the surface of his body subjected to torture, with ejaculation occurring when the pain was at its height – perhaps, like Lawrence, just before he passes out. In Anzieu's terms, Lawrence would be seeking to re-establish his skin-ego through his perverse masochism. However, Anzieu's interpretation of skin as having many functions allows us to add other skins to this understanding of Lawrence (if not Monsieur M).

There is, potentially, yet another (obverse) side to Lawrence's suffering – men's skins are often associated with both penetration by weapons and also flaying by whips (see Benthien 2002, pp. 63–94). In his recounting of the Deraa incident, Lawrence – probably knowingly – situates himself in a long iconography of male skinly privation, usually men who made great sacrifices for others, such as Christ, Saints and warriors. Undoubtedly, Lawrence's masculinity emerges from his relationship to his skin: a skin that must suffer; that must be penetrated; that must go through indescribable ordeals and trails; that becomes masculine at exactly the point that it is at its most vulnerable, its most penetrable, its most flayed. In this regard, Lawrence's sexual and physical need for John Bruce's birch become less associated with either homosexuality or homosociality, and less tied to masochism as such, but chained to a history both personal and quasi-religious that sees the suffering of the body and soul as a form of purity and redemption.

For Lawrence, then, the skin of suffering is never quite one thing: not simply masochistic, not simply homosexual, not simply heroic or anti-heroic, and not simply (Catholic) Christian. Fortuitously, for Lawrence, the war offered him a 'perfect storm' for his skin of suffering: indeed, we might wryly observe that what Monsieur M really lacked wasn't therapy but a jolly good war. Lawrence was never quite as on the edge of madness as he was during the war, never quite able to suffer as much, nor be as irrevocably lost. Years after the war, quietly settling into a secret life of anonymity and beatings, Lawrence hides from his skin, his self and the world: military service providing the perfect cover, another skin, for his skin of suffering.

Conclusion: The Chafing of Bodily Regimes

In the work of Freud, Bick and Anzieu, the space of skin and ego is a surface: skin as the surface space of the body; ego as the surface space of the mind; each a layer with two sides: one facing inside; the other facing outside. There are, however, two attendant dangers with assuming the space of skin and ego is on the surface. First, following the model of skin, it can be assumed that the surface of the ego is singular. Second, consequently, also following the model of skin, where gaps or holes or tears are found in the surface of the ego, that repair requires establishing a spatially integrated and coherent wrap (sometimes literally, see Anzieu 1985, pp. 120–122).

Instead, psychoanalytic work on the surfaces of skin and ego points towards the sheer variability of volumes, shapes and consistencies of skin and ego. Instead of a normative assumption about the space of skin and ego being a skin-like covering, we find a three-dimensional homunculus. In this vein, Lawrence of Arabia's experiences demonstrate how the skins of skin, second skin and ego can fail as a wrap or container. However, Lawrence of Arabia also shows not only how this failure can be inhabited as a skin-ego, but how the chafing of his skins can provide a ground for his skin-ego – his skin of suffering.

Lawrence, of course, is in many ways typical of the Imperial British male, secure in his knowledge of his superiority, of his civilisation, of his education (following Said 1978; see also Stoler 2002). It is this bred-in superiority, indeed, that allows Lawrence to assume that he, amongst all others, is the right man to liberate the Arabs. This allows us to conclude, of Lawrence, that it is his unmarked whiteness, the invisibility of his skin, that allows him to easily change sides and to pass as both Arab and Imperial Officer (see also Ahmed 2000, ch. 6). But this is not at all the case. Lawrence wanted to feel at home amongst Arabs, but didn't and wasn't; nor was he at home in the military command structure of the British

Empire. Lawrence is contradictorily territorialised: both/neither British and/nor Arab. The disjuncture between Britishness and Arabness provide Lawrence with an ideal ground on which to establish a skin of suffering, precisely by finding himself between these opposed worlds. Within and between these worlds, Lawrence found rich possibilities for his suffering.

Lawrence tried to prove himself, physically, to the Arabs: through his ability to ride the camel, to wear their clothes and eat their food, to endure the monotony and heat of the desert (see Dawson 1994, pp. 185 and 204). These sufferings, endured on and through skin, become part of the fabric of his masculinity, his hardening, yet vulnerable, body. Even so, Lawrence's 'white and fresh' skin would never pass as Arab, not even Circassian. His crimson face would mark him out from the British Officers, who viewed his beliefs and judgment with suspicion. Although Lawrence's skins allow him the seeming freedom and privilege of movement between the two worlds – British and Arab – he is marked as 'out of place' in both.

Lawrence never felt comfortable in the 'second skin' of his British uniform; it literally never quite fitted, his trousers, for example, were always far too short. We know that militarised bodies are built out of many resources, including those that work upon skin (e.g. toughening up) and second skin (e.g. uniform). However, Lawrence's experiences caution against reading the military body as fully militarised or militarisable – indeed, it might be better to see the militarised body as a particularly dangerous, yet unrealisable, myth. Though he loved the 'second skin' of his Arab costume, this did not fit Lawrence either. Neither second skin was a patch on skin itself. Lawrence remained split between two cultures: Arab and British: unable to adopt either fully as an identity.

Lawrence was never at home in his (white) skin, his (mad) ego nor his (costumed, uniformed) second skins. Lawrence himself despairs at being between worlds. Even so, it is possible that Lawrence was actually at home in the suffering that the chafing between these worlds, British and Arab, afforded him – especially, as this worked its way out through his skinly identity: his skins of suffering, the failures of his second skins. The misrecognition in Cairo and the rape at Deraa, in this light, become sites at which Lawrence took pleasure in his sufferings, rather than events that mark the failure or breakdown of his self. Here, pleasure takes the form of pompously teasing the Military Police, but also Lawrence's confessions over his sexual enjoyment of being whipped, and also his fantasies of rape. Lawrence, perhaps, is remarkable not because he contained disaffections and dismantlings such as this, but because he lived through his discomforts and displeasures, as part and parcel of his skin of suffering. His skin, his skins, simply did not fit. Of this, Lawrence was acutely aware. At times, Lawrence himself despairs at the madness of it. Lawrence had, like many others before and since, to negotiate feeling out of place no matter where he was. So, I have argued, he found his sense of self in his ever-chafing skins.

Lawrence's experiences help us understand the relationship between bodily regimes, policing, proper places and the distribution of the sensible. We saw that bodily regimes are imposed, as it were, from above. In his case, there are both British and Arab bodily regimes, each with their own way of assigning bodies to proper places. This not only involves bodies themselves, and parts of bodies, such as skin and voice, but also other things associated with the body, especially clothing (a second skin), but we might also add food and drink (water being a special concern for Lawrence). We can also see that policing bodies takes a literal meaning. And that the contestation of that policing also relies upon the resources available to particular bodies inside and outside of any particular bodily regime. Yet, I have hinted that the conscious appreciation of bodily regimes and the peculiarities and absurdities of their enactment can be undercut by the unconscious dynamics that underlie people's experiences of them. This accords with, what Chapter 1 described as, the unconscious structurings of the sensible and their distribution. These are seen as mutable, multiple, inconsistent and dynamic. Thus, rather than seeing the indeterminacy, inversion and acting out of the Deraa incident in Lawrence's life as exceptional, we can understand it as telling us something about the ordinary ways that the clash of bodily regimes structures people's experience of them.

Lawrence's seemingly perverse pleasure in his sado-masochist fantasies and acts tells us that unconscious dynamics and pleasurable forms are connected in some way and form part of what must be understood as a bodily regime and the clash between them. In Chapter 4, we develop our understanding of this relationship by focusing upon the relationship between bodies, passionate forms and unconscious dynamics. That said, the spatial lens of the chapter moves from thinking about proper places for the body to the topologies of physical and psychic space. Thinking about proper place allows us to see the topographical relationships between bodies, where bodies are assigned to particular places and spaces and the borders between those places are policed in various ways. We have seen this play out in the experiences of Frantz Fanon, Nella Larsen and T. E. Lawrence. We learned that the movement of bodies can be understood topographically, as the movement from one place to another. We also saw that bodies can shift in social space without having to move in topographical space, as their assignment to a place comes into question. Coincidentally, this happens for Fanon and Lawrence while on trains, but it can just as easily happen in a living room, a hotel lobby or a café (as Larsen shows). Thus, moving space – moving spatially – does not always require movement from one topographically defined place to another. To understand both the distribution of the sensible and the ways that bodies are assigned to spaces and places, then, requires more than a topographical account of space and place. Thus, the next chapter thinks topologically about space, primarily using (as Anzieu does; see above) the Möbius strip as an analytical metaphor.

Chapter Four
Bodies, Affects and Their Passionate Forms: Animal Phobias and the Topologies of Bodily and Psychic Space

Introduction

As we saw in Chapter 1, Rancière usefully opens up the question of the aesthetic unconscious that underpins Freud's use of the Oedipus myth and therefore, by extension, much (if not all) of his psychoanalytic account of the subject. Thus, for Rancière, a particular – classical – distribution of the sensible underlies Freud's understanding of the subject. What attracts Freud to the Oedipus myth is not so much the breaking of the incest taboo as the way the characters enact their destiny while trying everything they can to avoid it, the way they desperately seek the truth while at the same time being unable to see the truth. These paradoxes, inversions, reversals, ironies lie at the heart of subjectivity for Freud. Whilst the subject might be understood as inherently flawed, for Freud, these contradictions (etc.) arise out of the subject's encounters with the world. Thus, the social prohibitions on patricide and incest create social conditions for Oedipus' tragedy. No prohibition, no tragedy (see *Game of Thrones*, where incest and all kinds of -cides go unpunished).

Critically, the subject can be largely or entirely unaware of its contradictions, both internal and social. Instead, the subject can experience them, and indeed perform them, as symptoms. It is this unwitting aspect of the psyche that comes to preoccupy Freud, as it is this that, he believes, makes his patients sick. Increasingly, Freud focuses upon the mechanisms of repression – the internalisation and self-policing of social prohibitions – and consequently (what we can call) the repressed

Bodies, Affects, Politics: The Clash of Bodily Regimes, First Edition. Steve Pile.
© 2021 Royal Geographical Society (with the Institute of British Geographers).
Published 2021 by John Wiley & Sons Ltd.

unconscious. For Freud, understanding repression became the crux of analysis, as repression provided the key to the locked box of the unconscious. However, it's not just mum and dad that fuck you up (as Philip Larkin puts it). Childhood traumas, both large (devastating) and small (commonplace), come in many other forms. Indeed, it is hard to overstate the potential for adult symptoms in the sheer diversity of childhood traumas.

This observation has significance for our understanding of the distribution of the sensible. The distribution of the sensible is not simply underlain by a singular aesthetic unconscious (thought of as an unwitting aesthetic regime). Instead, we can suggest the distribution of the sensible as interfering in, and being interfered with by, unconscious dynamics both social and personal. This idea that the distribution of the sensible might be unconscious and also structure our understanding of the world can be usefully converted into a problematic. Thus, we can ask in what ways the distribution of the sensible and unconscious processes might be in relation to one another.

In this chapter, we focus on two people's experiences of animal phobias, as reported by Freud in one of his proto-psychoanalytic case studies of hysteria (1895), Emmy von N., and his psychoanalytic case study, the Wolfman (1918). In these case studies, either Oedipus is nowhere to be found (Emmy von N.) or extremely marginal to the analysis (Wolfman). Rather, Freud is preoccupied by his patients' bodily symptoms and their imaginary, fantastic worlds. In Rancière's terms, we might say that the patients are laying out their own personal distribution of the sensible, specifically through their animal phobias. These phobias structure the patients' bodily regimes. Through their bodies, Freud's patients enacted and re-enacted their traumas. For me, there are also correlates with the model of the skin-ego laid out in Chapter 3.

In the discussion of T. E. Lawrence's *Seven Pillars of Wisdom*, I argued that the skin-ego is not one model for the development of the ego, but many. It is not simply that the skin has many functions, so cannot provide a single path down which the ego must travel. It is as much that these functions coexist, providing many paths, and that these divergent functions and paths can create models for the psyche-body that are in conflict with one another. More than this, the particular development of a skin-ego is a negotiation between the body-psyche and the world – or rather, as Fanon, Larsen and Lawrence demonstrate, worlds (plural). Thus, we should not see animal phobias as providing a singular, nor coherent, bodily regime for either Emmy von N. or the Wolfman.

Nor can we say that the unconscious structuring of the distribution of the senses produces a singular, or coherent, world of the senses. This again correlates with the model set out in the previous two chapters. Fanon and Lawrence, in particular, had to negotiate living with different worlds, with being both inside and outside in these different worlds. They adapted to, and actively sought, embodied lives that matched their desire for their own bodily regimes to match the bodily regimes of the world. As in Larsen's *Passing*, we have seen that understanding this movement between inside and outside is critical for understanding how people negotiate different bodily regimes.

So far, we have understood this inside and outside movement in two ways: on the one hand, through encounters between an individual and the world that lays bare the differences between internal and external worlds, as with Fanon; on the other hand, through the movement of people across the spatial and social boundaries that are designed to separate one bodily regime from another, as with T. E. Lawrence and Larsen. Thus far, inside and outside have been understood topographically, as distinct spaces (Lawrence in Arabia) and as distinct social spaces (through passing). Topography usefully privileges the construction of borders (of all kinds) between spaces and the policing of those borders (sometimes using actual police). However, we need also to understand how inside and outside are constantly formed and reformed, relationally, and how they change shape and form (often in ways that are consistent over time: as in the changing same of the racialisation of space, for example). This requires us to think topologically about inside-and-outside (hyphens added: a unity comprised of opposites). This is significant because bodily regimes work both topographically and topologically. As does, I argue, the distribution of the senses.

In this chapter, we focus on the analogy of the Möbius strip to understand the spaces of inside and outside. The Möbius strip is (famously) formed by taking a strip (of, say, paper), giving it a twist, then joining the ends together. There are three features of the Möbius strip that are important. First, there is always in inside and an outside. However, this inside and outside is always relational. So, analogically, we can use this idea to think about the ways that the outside of skin (etc.) can reveal or create different kinds of insides, such as in blushing or sweating, pigmentation or scarring. Second, the flipping of inside and outside involves, in the Möbius strip analogy, movement along a path. This tracks the movement of physical and psychical material along whatever boundary defines the inside and the outside. Indeed, boundary may flip the inside and outside while at the same time being itself fixed and static. Third, and perhaps the least well utilised aspect of the Möbius strip, is that topological form is not just a surface, it is also a volume, in 3D. This means that there has to be the possibility of lateral movement across the strip, which enables a change of relative location, without changing the relationship between inside and outside. This movement across enables us to think about change in the (psychical, social, physical) location of the subject without a change in relationship between the subject and the inside/outside. In this chapter, we focus on how these three kinds of movement – along, over and across – create meaning many times over: that is, produce overdetermination.

To highlight this movement along, over or across a path that is relationally overdetermined, yet constantly indeterminate, we will take a close look at two of Freud's case studies, Emmy von N. and the Wolfman. In both cases, we will follow the animals through the body and the psyche. This allows us to see that bodily regimes are built up out of conventions of human worlds (what Rancière 2004, might call either a 'community' or a 'distribution of the senses') that rely upon a distinction between the human and the beastly.

Animal Tracks Inside and Outside the Body and the Psyche

The focus of this chapter is on how animals can become a model for the body-psyche (like skin did for the skin-ego in the last chapter), yet in ways that reveal the movement between inside and outside and the difficulties of, and solutions to, being both animal and human. I have chosen animals, partly because they shed new light on the development of Freud's own understanding of the relationship between unconscious processes and the body, but also because commonplace understandings of animals and humans tends to place them poles apart (especially, ironically, when these poles are glued together as a hybrid, as in iterations of neologisms such as natureculture). To achieve this, I explore two of Freud's case studies: Emmy von N. (Freud 1895) and the Wolfman (Freud 1918).

Both Emmy von N. and the Wolfman experience a fear of animals, each living and performing their fears in markedly different ways. However, what they share is a close connection between fear, desire and the animal. Freud, of course, is seeking to understand these connections clinically so that he can intervene in them therapeutically. What this connection produces, for both Emmy von N. and the Wolfman is an animal form for their passions, both desires and fears. This reveals that desires and fears require forms to be experienced, enacted and performed. Animals, of course, are far from the only form that the passions can take. It is easy to think of others, but it is important to include all kinds of objects, not just bodies or body parts, not just individuals or peoples, but also ideas, acts, practices and even politics. Indeed, one of Freud's therapeutic errors, at least initially, was to privilege these other forms before he started to take the animals seriously. In his attempts to ameliorate Emmy von N.'s hysterical symptoms, Freud pays little attention to the animal stories of his patient, and his patient and her therapy suffers as a consequence. In his therapeutic interventions with the Wolfman, Freud has learned to pay close attention to his patient's animal phobias, with direct consequences for therapeutic practice and success.

Eventually, Freud comes to realise is that animals are a *passionate form* for his patients, either giving body to an ever-evolving set of phobias (Emmy von N.) or to fetishistic desire (Wolfman). Here, I use the term *passionate form* for two reasons. First, I am suggesting that the passions (fears, desires, anxieties) need a form to find expression. This form does not need to be consciously expressed – and can be expressed in all kinds of ways. This is to see, for example, the expression of hysterical symptoms as a form for the passions as much as more obvious forms such as kissing or fetishes, or indeed more rarefied forms such as fine art – and politics. Second, what this term captures is the necessary entanglement of passion and form: without a form passion cannot find expression, while the passions are always in search of an appropriate form through which to express themselves. In saying this, I am giving passions a certain agency. They are dynamic, mutable and multiple. However, the important thing here is that the passions can shift from one form to another, sometimes easily, sometimes rapidly, often inexplicably. This we will see in the mobility and proliferation of Emmy von N.'s hysterical symptoms.

To understand how the passions can stay the same while also finding ever new forms is usefully explored through a topological analysis. In particular, it is worthwhile drawing on one topological figure, the Möbius strip; its significance for psychoanalytic theory originally identified and explored by Lacan (see 1959). Topologically, the strip converts two sides into a single surface, through twisting and joining. The Möbius strip, at one and the same time, has both one side and two sides. For Lacan, this provides an excellent metaphor-and-ontology for psychical processes that are capable of holding contradictions (two opposing ideas) together without contradiction (as a single idea). In this way, the strip echoes Freud's suggestion that the unconscious is able to hold contradictory ideas (such as desire for, and disgust of, an individual) without contradiction. But Lacan was also thinking more abstractly about how dyads, such as the mind-body or conscious-unconscious, are constantly undone and redone (see Wegener 2016; and Cohen 2016). Such thinking allows us to expand the significance of the Möbius strip.

There is a specific advantage of the Möbius strip over, say, the torus (donut) or Klein bottle. The torus and Klein bottle are often used to describe the ways that insides become outsides (and vice versa) (for an illuminating discussion of topological analyses, see Allen 2016; see also Martin and Secor 2013). For John Allen (2011), the spatial foldings and inversions in these topological figures enables him to see how power relationships utilise space in ways that do not require control over territories or through networks. However, the specific advantage of the Möbius strip is that its twist provides an analogy for a change in space, from one form to another. This twisted topological thinking enables us to see how the human and the other-than-human are always already 'in relation', yet also that this can be transformed in various ways while leaving the original relationship intact. To explore this paradox further, I draw upon the foundational work on the topologies of the psyche by Virginia Blum and Anna Secor (2011).

Blum and Secor explore the significance of topological thinking for understanding the relationship between psychic and material space. They argue that Freud's topographical model of the psyche (1923), grounded in a territorialisation of different functions of the psyche (conscious, preconscious, unconscious), must be supplemented by a topological understanding (drawing upon Lacan 1966 and 1973). While a topographical understanding of space can illuminate its 'mappable, graphable and measurable' aspects (2011, p. 1034), this does exhaust how space works. On the other hand:

Topology deals with surfaces and their properties, their boundedness, orientability, decomposition, and connectivity – that is, sets of properties that retain their relationships under processes of transformation. (p. 1034)

There are clear links between the idea of boundedness and surface and the notion of a skin-ego, albeit conditionally and contextually. We can also see that central of the use of topology is the idea of transformation: that is, change in form. Thus, topology captures a change in form, while the relationship that comprise that form remain the same. This spatialises the commonplace expression, 'the changing same'. Thus, topological thinking becomes useful for thinking about how passionate forms can undergo processes of transformation, yet retain their original relationships. However, for me, topologies must themselves be supplemented by topographical thinking, as this allows us to see the processes that prevent transformation. As importantly, relationships can change even while forms remain the same. Thus, topological and topographical thinking proceed hand in hand. Indeed, arguably, a topography is merely a fixed topology, where processes of transformation have been successfully arrested or indeed where they are absent or have been resolved (however temporarily).

The following sections emphasise the appearance of animals in Freud's patients' symptoms. Intriguingly, although each patent's symptoms seemingly points to a different pathology – hysteria (Emmy von N.) and neurosis (Wolfman) – animals loom large. To begin with, I briefly outline two of Freud's case studies. In each case study, Freud's repeated (and often futile) attempts to understand the causes of symptoms are clearly important, as are some of his therapeutic interventions, in clarifying the role of animals. Revisiting these case studies with animals in mind enables me to rethink the interaction between humans and objects *psychoanalytically* – that is, both dynamically and unconsciously. This reveals that clear demarcations and dispositions of identity become increasingly fuzzy and indeterminate the closer you get to them.

I am concerned to learn from Freud's shifting attention to the agency of animals in his therapeutic practice. In this shift, we can begin to discern Freud's break from orthodox representations and diagnoses of hysteria, away from visual representations (in drawings and photographs, especially) and detailed verbal descriptions written by clinicians, towards accounts of their symptoms in patient's own words (see de Marneffe 1991). In many ways, this confirms Rancière's (2001) suspicion that Freud was unconsciously deploying an aesthetic regime based on character flaws, as these are revealed in the gaps between deeds and words. Thus, the representational regime becomes inadequate: symptoms are not things in themselves, they have other meanings; their meanings determined, many times over.

Let us start with Freud's treatment of Emmy von N., the earlier of the two case studies.

Do Not Touch Me! Emmy von N.'s Fear of Mice and Other Animals

Much is now known about Emmy von N. and her family context (see, e.g., Appiganesi and Forrester 2005); in addition, there has been much debate about the course of Freud's treatment of Emmy von N., though the psychoanalytic literature has focused almost exclusively on (re)diagnosing her psychological condition and the implications of this for current psychotherapeutic practice (see, e.g., essays in Dimen and Harris 2001). Here, I focus on the appearance of animals, and their consequences, in the course of Freud's analysis and treatment of Frau Emmy.

Freud's outline of his treatment of Emmy von N. is presented chronologically, as if lifted directly from his notebooks. The case study unfolds as a series of difficult encounters between Freud and his patient, during which Freud eventually manages to alleviate Emmy's suffering – although not completely, and not without a battle of wills between the pair (Campbell and Pile 2015). As yet, Freud is not a psychoanalyst. Instead, following the lead of Jean-Marie Charcot and Josef Breuer, his faith lies in the therapeutic value of hypnosis (see Pappenheim 1980). The case study describes Freud's (increasingly frustrated) use of hypnosis to uncover the secret traumas that lie behind Emmy von N.'s hysterical symptoms, as we will now see.

On 1 May 1889, Emmy von N. puts herself in the care of Freud, after a recommendation from Josef Breuer (Ellenberger 1977). She was, Freud confidently declares, 'a hysteric'. He observes:

I find a woman who has distinctive, finely cut features and an appearance that is still youthful lying on the couch with a leather bolster under her neck. Her face bears a tense, pained expression; her eyes are screwed up and cast down; she has a heavy frown and deep naso-labial folds. She speaks as if it were arduous, in a quiet voice that is occasionally interrupted to the point of stuttering by spastic breaks in her speech. When she speaks she keeps her fingers, which exhibit a ceaseless agitation resembling *achetosis* [involuntary writhing movements of the fingers, limbs and neck], tightly interlaced. Numerous tic-like twitches in her face and neck muscles, some of which, in particular the right *sternocleido-mastoid* [a muscle running down the neck, from below the ear to near the collar bone], protrude quite prominently. In addition, she frequently interrupts herself in order to produce a peculiar clicking noise, which I am unable to reproduce. (1895, pp. 51–52)

In a footnote, Freud adds:

This clicking proceeded as a number of tempi. On hearing it, colleagues of mine who knew something about hunting compared the final notes with the mating cry of the capercaille [heather cock]. (p. 96)

Indeed, please note, the heather cock is renowned for its mating display. Freud continues his description of Frau Emmy. She is intelligent and educated. But this only makes Freud more uncomfortable with her symptoms. It is:

> all the more disconcerting that every few minutes she should suddenly break off, contort her face into an expression of horror and revulsion, stretch out her hand towards me with her fingers splayed and crooked, and in an altered, fearful voice call out the words: 'Keep still – don't say anything – don't touch me'! She is probably under the influence of a recurring and horrible hallucination and is using a formula to fend off the interference of a stranger. (p. 52)

Already, even in their first meeting, Freud becomes all too aware of Emmy von N.'s long history of hysterical symptoms and their (unsuccessful) treatment, especially since the death of her husband fourteen years previously (in 1875).

The following day, 2 May 1889, Freud sets out a regime of analysis and treatment. In the morning, Frau Emmy receives a warm bath and a whole body massage (p. 53). The massage is given by Freud, who has plainly decided to ignore Frau Emmy's instruction not to touch her. Indeed, the treatment of a whole body massage is prescribed twice daily. In the evening, Frau Emmy undergoes a hypnotic session with Freud. 'She is perfectly suited to hypnosis', Freud observes, 'I simply hold a finger before her eyes, exclaim "Sleep!", and she sinks back looking dazed and confused' (p. 53).

In these sessions, Freud uses the technique of *suggestion* to influence Emmy von N. upon waking from her trance. The therapeutic use of suggestion during hypnosis was common in the late nineteenth century. The basic idea was to seed ideas unconsciously that would cure the patient of their symptoms (see Pile 2010). Although Freud does not say what his hypnotic suggestions are, it is likely that he would discuss each symptom with Emmy, then make the suggestion that, on waking from the hypnotic trance, she would no longer suffer from it – employing the same technique that Josef Breuer had, seemingly successfully, used with Anna O. (Breuer 1895). Freud is initially pleased with the results, as Emmy seems very responsive to his suggestions. Over the course of the first week of treatment, her condition visibly improves.

Yet, on 8 May, Emmy unexpectedly 'entertains [Freud], in what seems to be a normal way, with some horrible animal stories' (p. 53). In particular, Emmy draws Freud's attention to a story she's seen in the *Frankfurter Zeitung*. According to Emmy, 'an apprentice had tied up a boy and put a white mouse in his mouth'. 'The boy', she continued, 'had died of shock' (p. 54). She then told Freud that Dr K. had sent a box of white rats to Tiflis (Tbilisi). Freud recalls how Emmy's body recoiled in horror at the idea: her hands clenched and unclenched, she visibly quivered. 'Keep still, don't say anything, don't touch me'!, she exclaims. Tormented, Emmy, still clearly thinking about the package of rats, fantasises:

> If an animal like that were in my bed! (Shudders.) Just think, when it's unpacked! There's a dead rat among them, a rat that has been gn-aw-ed at. (p. 54)

Freud's response takes two forms. On the one hand, he attempts to relieve Emmy from her 'animal hallucinations' using suggestion in the evening's hypnotic session. On the other hand, Freud tracks down the story of the apprentice in the *Frankfurter Zeitung*. Intriguingly, he discovers that the boy's mistreatment involves neither mice nor rats. He surmises that the story must have prompted a delirium, during which Emmy *added* the white mouse to the story. In the evening, strangely, Emmy cannot recall her conversation with Freud about the mice and rats. Indeed, she is astonished and laughs at its absurdity.

Later, during the hypnotic session, Freud decides to press Emmy on her animal stories. Why, he wonders, was she so easily frightened? Replying, Emmy conjures up some early childhood memories. 'First when I was five and my brothers and sisters threw dead animals at me so often, that was the first time I fainted and had convulsions' (p. 54). Emmy recalls, in quick succession, other traumatic experiences, each to do with death. At seven, she sees the body of her sister in her coffin. At eight, her brother frightens her by dressing up as a ghost. At nine, she is looking at the dead body of her aunt, when suddenly the jaw falls open. Emmy is clearly disturbed as she tells Freud these stories. Finally, Emmy opens her own mouth wide and gasps for air. Then, she becomes peaceful.

Freud's reaction is to attempt to erase these images of animals and corpses, using hypnotic suggestion. To reinforce the suggestion, Freud passes his fingers over Emmy's eyes, so her eyes will be permanently closed to the animal hallucinations and images of death. It does not work. The following day, Emmy is suffering from another hysterical attack, this time prompted by an image of 'Indians dressed as animals' (p. 55) in an atlas her daughters have shown her. Freud, in the evening's hypnotic session, wonders (again) why this was so frightening? It reminds Emmy of her brother's death. This memory conjures up another memory, of a serious illness her sister suffered. Freud once again turns to suggestion as a solution: next time she sees the picture of the Indians, Freud tells her, Emmy will not be frightened of them; instead, she will laugh. By the evening, Emmy is in very good spirits – good enough, even, to make fun of her treatment (which does not amuse Freud).

Emmy's fears seem to be the key to her hysterical symptoms. In pursuit of an understanding of these fears, Freud asks Emmy for further traumatic experiences. What he does not do, however, is follow Emmy's fear of animals. This probably does not seem strange. After all, Emmy's hysteria is obviously connected to her fear of death and dead bodies. However, Breuer's technique, which Freud has been following closely, involves excavating earlier and earlier experiences. Indeed, in the case of Anna O., Breuer had shown that early childhood experiences turn out to be foundational in the formation of hysterical symptoms. Freud ought to have followed the mice. He didn't.

Instead, Freud discovers that, when she was 15, Emmy found her mother lying on the floor, having suffered a stroke. Four years later, Emmy came home to find her mother dead. She was 19. Curiously, though, this memory prompts another from the same time. Emmy found a toad underneath a stone: she was unable to speak for hours afterwards. Freud responds like a fire-fighter putting out a wildfire: he seeks to extinguish all these memories one by one – stricken mother, dead mother, toad, being struck dumb and all. As before, it does not work. Days later, Emmy is haunted by nightmarish dreams – 'the arms and legs of chairs were all snakes; a monster with a vulture's beak stared at her; other wild animals leapt at her' (p. 63). Then, she recalls reaching for a ball of wool, only for it to be a mouse that then runs away; then, seeing a toad while on a walk, which suddenly sprang at her.

Freud's attempts to extinguish Emmy's memories are proving futile: the wildfires keep breaking out in other places. Emmy, too, realises this and becomes suddenly pessimistic about the prospect for a cure. Yet, on 13 May, Emmy thinks she has made a breakthrough. She realises that small animals loom so large for her because of a stage act she'd once seen: during the performance, an enormously large lizard appeared. Unfortunately, this realisation does not really help Emmy.

Over the next few days, the familiar pattern of recollection and erasure continues, as Freud seeks to smother the wildfires of Emmy's hysteria. It fails. Emmy's animal horror stories continue unabated: a toad in a cellar (p. 67), mice sitting in trees, the hooves of horses at the circus, the mouse she mistook for a ball of wool (again), being chased by a bull, twice (all p. 69). Freud forbids Emmy to remember. It does not work. The fires will not go out. Emmy recalls discovering hundreds of tiny worms in bran; a bat caught in her wardrobe; a bat-shaped broach she can never bring herself to wear; a pathway in St. Petersburg so covered in toads they could not walk on it (all p. 70). Yet, Freud persists. He asks Emmy to list the animals she is afraid of, and tells her one by one not to be frightened of it. Emmy determinedly wills herself to respond: 'I mustn't be afraid' (p. 70).

Rather than worry that the technique is not working, Freud is anxious that he is being neither exhaustive nor thorough enough. By May 17, both Emmy and Freud appear at their wits end: Emmy is losing her patience; Freud is losing his patient. Yet, even though exhausted and frustrated, Emmy's hysterical symptoms have been alleviated. Indeed, the very next day, 18 May, sees the last of the serious attacks. Just four weeks later, Emmy is discharged and returns home. She is well for a while, but seven months after that Emmy is once again in the consulting room of a doctor (not Freud), seeking hypnotic therapy.

By focusing on Emmy's horror of animals, it is possible to see a couple of things not usually evident. First, let me say what is evident. Freud's 'cure', using hypnotic suggestion, is only partially successful and only temporary. Freud will later abandon the use of hypnosis and hypnotic suggestion in his therapeutic practice and, in its place, create a new form of therapy: psychoanalysis. Yet, in the course of Freud's treatment, we can now glimpse how Emmy repeatedly attempts to convert Freud's treatment into something more psychoanalytic (see also Miller 2012): she wants him to pay attention to her, to listen to her animal stories, to follow the connections between those stories, and map out the underlying traumas that provide these stories with their emotional and affective intensity. Despite Emmy's clear wish that he should follow her animal stories, Freud does not do so. Instead, he seeks to stamp them out. At every turn, he instructs Emmy to forget. However, the zoo of Emmy's mind will not simply be erased. First, there are mice and rats; then, there are horses and toads; then, there are bulls and bats. Freud has not yet understood how dynamic the unconscious is, nor how it seeks to represent itself – in this case, through an ever-expanding menagerie of animal horrors. In the next case study (the Wolfman, below), Freud shows that he has learnt Emmy's lesson. He will seek to place the animal stories in the context of familial relationships. So, ironically, one reason for the failure of Freud's treatment of Emmy is that he does not follow Breuer's technique closely enough: he does not track Emmy's animals into her earliest childhood experiences. Consequently, Freud is unable to see the beasts in Emmy's mind as either dynamic or constitutive of her psychical structures.

Freud remains only half-aware that Emmy's behaviour is, in part, modelled on the animal: thus, it is only in a footnote that he notes that her strange vocal sounds might mimic the mating behaviour of a bird – and he certainly doesn't use this observation to consider Emmy's sexuality. He fails to inquire about the syncretic animal-human figures that terrify Emmy. Indeed, a large part of Emmy's horror appears to concern the incorporation of the animal into the human, whether by mimicry or by eating or by touch. We can observe at this point that Emmy's psychic structures do not simply 'house' a zoo of animal stories, nor does her mind simply conjure up animal hallucinations as if these were somehow external objects. Rather, these animals make Emmy who she is and how she behaves: her symptoms do not form around animals, but through the animal, and this is why her animal horrors proliferate despite Freud's best efforts to stop them in their tracks. We might say Emmy's humanness is already beastly. But Freud cannot, yet, see this.

Indeed, invisible to Freud is an aesthetic unconscious that is motivated by, and circulates around, people's repressed encounters with, and thoughts and feelings about, animals. These thoughts and feelings are not simply divisible into categories of animal and human (i.e. a polarised distribution of the sensible), they lie along a continuum that enables the reversal of the human into the animal and the human into the animal (i.e. a looped distribution of the sensible). Indeed, even the name Wolfman appears express this unity of opposites (yet another kind of distribution of the sensible), by acting as a concatenation of the wolf and the man; thereby, embodying the jointing of the animal and the human. However, it is this jointing, this movement along a continuum, this experience of unity of opposites that requires explanation. To understand the beastliness of the human a little more, I now turn to Freud's study of S.P. (written after the conclusion of treatment in Winter 1913/1914, but not published until 1918). Here, Freud does track the animals, most famously the wolves.

More Ferarum: More than a Wolfman

Titled as a study of infantile neuroses, and therefore about the lingering experiences of childhood in human life, Freud notes that his patient's initials are S.P. (1918, p. 293): short for Sergie Pankeyev (25 December 1886 – 7 May 1979). However, rather than using a pseudonym (as with other case studies), this case study is most widely known by its hybrid animal-human name: the Wolfman – a creature of mythic proportions. Too mythic. And too hybrid. S.P. is, unfortunately, *over-identified* with the wolf. It is better to think of S.P.'s dream of the wolves as a turning point in the analysis, rather than assuming that S.P.'s neuroses were governed by his 'wolf-man-ness'. Indeed, S.P.'s neuroses were long-standing and took a variety of forms through his life. Freud summarises:

> The case concerns a young man who suffered a physical collapse in his eighteenth year following a gonorrhoeal infection; when, several years later, he came to me for psychoanalytic treatment he was completely dependent and incapable of autonomous existence. He had lived more or less normally during the decade of his youth that preceded the illness and had completed his secondary school studies without undue disruption. His earlier years, however, had been dominated by a serious neurotic disorder which began shortly before his fourth birthday as anxiety hysteria (animal phobia) and then turned into an obsessive-compulsive neurosis [*Zwangsneurose*], religious in content, the ramifications of which persisted until his tenth year. (Freud 1918, p. 205)

As with Frau Emmy von N., a great deal is now known about S.P. and his life after treatment by Freud (see, e.g., Obholzer, 1980; see also Campbell and Pile 2011). As with Emmy, my focus is on the animals in S.P.'s life. Here, it is important to note that more than just wolves occupy S.P.'s mind. But, in Freud's analysis, an understanding of S.P.'s beastly mind begins with his dream of the wolves.

On the night of Christmas Eve, just before his fourth birthday (which was on Christmas day), young S.P. has a nightmare:

> *I dreamed that it is night and that I am lying in my bed (the foot of my bed was under the window, and outside the window there was a row of old walnut trees. I know that it was winter in my dream, and night-time.) Suddenly the window opens of its own accord and* terrified. *I see that there are a number of white wolves sitting in the big walnut tree outside the window. There were six or seven of them. The wolves were white all over and looked more like foxes or sheepdogs because they had big tails like foxes and their ears were pricked up like dogs watching something. Obviously fearful that the wolves were going to gobble me up I screamed* and woke up. My nurse hurried to my bedside to see what had happened. It was some time before I could be convinced that it had only been a dream, because the image of the window opening and the wolves sitting in the tree was so clear and lifelike. Eventually I calmed down, feeling as if I had been liberated from danger, and went back to sleep. (Freud 1918, p. 227: the dream part of S.P.'s recollection is in italics in the original)

Already we can see that the dream is about more than simply wolves. Obviously, the fear of being gobbled up by the wolves is the central terror represented in the dream. But the wolves are not simply wolves – they are also Russian silver foxes because they have big tails, their ears are pricked up like dogs, and they are completely white, so they more closely resemble foxes or sheepdogs than the darker Eurasian wolf. Perhaps, the Wolfman is *also* the Foxman or the Dogman (as Haraway might attest, 2008, pp. 28–30).

Freud's practice of analysis and treatment has moved on since his experiments with hypnosis and suggestion. Now, his method is psychoanalytic; its central technique is dream analysis (see Freud 1900; see also Davis 1995; or, for a different application of the method, Pile 2005a). Freud's method remains thorough and exhaustive. He will track down every association that S.P. can make to each dream symbol. . .and then he will follow S.P.'s associations makes with these stories. . .and, by following the extended networks of associations, Freud will painstakingly build up a picture of the hinterland of affect and meaning that the dream connects to. Instead of ignoring the animals, Freud wants to know more and more about them. Why wolves? Why white? Why foxes? Why sheepdogs? Why is the window the only active dream symbol? And what about the walnut tree? And so on.

A dense and intense set of images and stories emerge slowly from S.P.'s associations. As with Emmy von N., significantly, these associations lead to some horrid animal stories. Thus, as you might expect, Freud wonders about whether the dream connects to Little Red Riding Hood's encounter with a wolf. S.P., however, makes a link to the story of 'The Wolf and the Seven Little Kids' (for a contemporary rendering, see Hoffman 2000). In the story, a nanny goat has seven kids. While the nanny goat is away, a wolf tries to trick the kids by pretending to be their mother. At first, they see through the disguise. The wolf tries again, talking to them through a window, while standing on his hind legs stretching to reach the window ledge. Not to be fooled, the children spot the wolf's hairy feet. But the sneaky wolf disguises his feet in white flour, so when he puts them on the window ledge a second time, the children are kidded into thinking it is their mother. They let the wolf in. On seeing the wolf, the children scatter. But the wolf finds them, one by one, gobbling them all up in one bite: all, that is, except one. When the nanny goat gets back, she finds carnage. She is so grief-struck that she wanders out of the house into a near-by meadow, with the last remaining kid trailing behind. There, they find the wolf fast asleep. The kid is sent to fetch a knife. They cut open the wolf. Miraculously, the six swallowed kids are still alive. They replace the kids with stones and carefully sew up the wolf's stomach. The stone-stuffed wolf eventually drowns. Phew!

Certain elements from the story have been re-used in S.P.'s dream: the wolves are made white, as if covered in flour; there are six or seven wolves in the dream and seven kids in the story, six of whom are eaten; in the dream and the story, a window acts to separate the world inside from the world outside, yet the threatening outside world can also be seen, and potentially get in, through the window. In both the dream and story, the wolf threatens to gobble you up. Indeed, kids can refer to human and animal offspring as well as both at the same time; significantly, the same can be said about nanny, except combining nursemaid and goat. These ambiguities prove highly significant during the analysis. Intriguingly, Freud also discovers that S.P.'s grandfather is also fond of telling his grandchildren, in jest, that he will cut open their tummies. Not seeing the joke, this actually frightens them. Worse, S.P.'s father teases his kids in a similar way, telling them that he will gobble them up. Consequently, Freud begins to wonder whether S.P. has come to see the wolf in story of the seven little kids as being like his father – and vice versa (1918, p. 230). The Wolf-man is child to a Wolf-father.

The animal stories are not over, however. S.P. recalls the tale of a fox that tries to catch fish with its tail, only to have the tail break off when the water freezes. This leads to another tale. His grandfather told him the story of the wolf and the tailor. The tailor is sitting in his room, when a grey wolf jumps through the window. The tailor bravely grabs the wolf by his tail. The tail comes off in his hand and the wolf runs away. Later, while walking in the forest, the tailor happens upon a pack of wolves. He climbs into a tree and sits motionless. But the maimed wolf is amongst them and proposes to reach the tailor by forming a pyramid. He suggests the other wolves climb upon him. The tailor spots the tail-less wolf and shouts 'Catch the grey one by its tail'! The wolf runs in terror, causing the pyramid to collapse. Two elements of this tale are key to Freud's interpretation of S.P.'s dream (see Freud 1918, pp. 233–234; and also Pavda 2005). First, there is the symbolic castration of the wolf by the tailor pulling off its tail. Second, there is a pyramid of wolves, in which one wolf climbs on another's back.

Two more animal stories are needed before Freud feels confident that he has a full understanding of the dream. *First*, the symbolism of whiteness is about more than flour. For S.P., whiteness connects the whiteness to flocks of sheep and the sheepdogs on his father's estate. Whiteness is also associated both with bed linen and with his parents' bedclothes. Consequently, whiteness becomes a symbol for sexual exploration. It is, also, connected to death. *Second*, S.P. has a history of tormenting animals. He took pleasure in pulling off the wings of flies, stomping on beetles, and, in his imagination, beating large horses – all to annoy his nanny, Nanja (Freud 1918, p. 224). Indeed, in a later dream, S.P. sees a man tearing off the wings of an asp – an asp, Freud determines, being a mutilated wasp (1989, pp. 292–293). Perhaps unsurprisingly, S.P. is also afraid of a big yellow butterfly, which he has chased; he is disgusted by beetles and caterpillars; and also screams if he sees horses being beaten (Freud 1989, pp. 213–214).

What Freud can see in the Wolfman, *that he was unable to see in Emmy von N.*, is the extent to which animals create an intense and ambivalent 'primal scene' for S.P. Although shrouded in doubt, speculation and fantasy, Freud and his patient together uncover the probability that much of S.P.'s early sexual and erotic experiences are determined by his observations of animals, by his projection of these onto his primary 'love objects' – that is, his father, his mother, his sister and his nanny – and by his assumption that his love objects behave *in the manner of animals*. Some of these ambiguous human-animal relationships are alluded to in the dream, where the six or seven wolves might refer to some or all of S.P.'s love objects (see Campbell and Pile 2011). But, S.P. has one more story to tell. Sometime between the age of 18 months and 4 years old, perhaps when he was three and a half, S.P. remembers – or perhaps fantasises? – walking into his parent's bedroom and seeing them having sex. Freud reasons:

> In my opinion, it is possible to interpret the facts of the case as follows: we cannot forgo our assumption that the child observes coitus and in doing so acquires the conviction that castration might be more than an empty threat; the significance adhering to the positions of man and woman, in the first place for the development of his fears, and subsequently as a condition of intercourse, leaves us no choice, moreover, but to conclude that it must have been 'coitus a tergo' [from behind], 'more ferarum' [in the manner of beasts]. Another factor is less crucial, however, and could be left aside. The child might have observed coitus between animals [i.e. sheep, dogs and/or horses], rather than between his parents, and then imputed it to his parents, as if he had decided that his parents would not do it any other way. (1918, p. 256)

Or, put another way, Freud and S.P. have no idea whether he saw his parents doing it 'doggy style' (to use a popular contemporary expression) or not, but what they are sure about is the gluing together of two ideas: sex between his parents and sex in the manner of beasts. Whatever its reality, S.P.'s primal scene is founded on animal-sex. This single idea – sex from behind in the manner of beasts – has spiralling consequences for S.P. and his developing sexuality. For our purposes, the important point is not about castration, nor S.P.'s erotic fantasies about, or experiences of, his father, mother, nanny, maid and sister – the shifts and inversions of which being the traditional focus of the psychoanalytic literature (starting with Freud). Instead, it is about the *nature* of the primal scene.

As with Emmy von N., S.P.'s animal phobias would continue into later life. He would report anxiety dreams that carried images of lions, horses, a giant snail and being pursued by a giant caterpillar (Freud 1918, p. 268). What is important is the entangling of the human and the animal: animals represent people, yet they are also stories about those animals told by people, often as warnings about the behaviour of people – be it in the form of the greedy wolf, the sly fox, the stupid pig or the lucky kid. What the Wolfman shows us, which we can only suppose is also the case for Emmy von N., is that the 'primal scene' is always already an intense, ambiguous drama, involving people and animals, fantasies and theories. In this scene, animals are not simply representations of people, but a concatenation of often contradictory ideas: often involving people, but also involving relationships between animals (e.g. foxes and fish, or goats and wolves), even as these also stand for relationships between people (e.g. the hunter and the hunted, the trickster and the dupe).

Further, it would not help to make a distinction between the human and the animal (and other objects) in the course of psychoanalysing the Wolfman, for precisely what Freud is attempting to follow is the twists and turns of affect and meaning, as these travel through S.P.'s entangled ideas about people and things. Whether the Father is actually like a wolf or not is not Freud's question. Instead, Freud is more concerned about when the father appears as a wolf, about S.P.'s sense of his father's wolf-ness, and what this means for him. As a result, Freud can no longer distinguish, precisely, between man and beast – so it is hardly surprising that S.P. acquires, over time, the nickname Wolfman. Perhaps surprisingly, Freud himself does not refer to S.P. as the Wolfman in his essay. Indeed, the original 1918 essay title does not even nickname the case study 'The Wolfman'. A similar observation can be made about Freud's other great hybrid case study, the Ratman (1909) (see also Blum and Secor 2011).

What the hybrid does is provide a point of intersection between the animal and the human: the Wolfman, the Ratman. This intersection appears to be a single point where two lines identity construction cross, seemingly consequently providing a clear identity. Yet, this intersection is determined many times over, as 'man' and 'wolf' undergo interpretation both conscious and unconscious. It is not just that the point of intersection is unstable as a consequence of their meeting, but that the lines of identity are themselves fuzzy. The overdetermination of meaning of wolf and man itself producing indeterminacy. We have thought about this through the idea of the Möbius strip. The strip is important, in this context. It is not a Möbius line, but rather a volume in three dimensions. This means there is lateral movement across the strip, the accumulation of which creates a fuzziness. More than this, there's more than one point of intersection. There are many animals, many others, many encounters. Each of these affords an opportunity for the production of passionate forms, prompted by the unconscious structurings of distributions of the sensible.

From the All Too Human. . .to the Animal Human

Mouse. Wolf. Rats. Bats. Butterfly. Sheep. Dogs. Foxes. Nanny Goat. Kids. A little boy. Little girls. Nanny. Mother. Heather cock. Father. Tailor. Wolves. Beetles. Brothers. Sisters. Caterpillars. Flies. Worms. Horses. A bull. A large lizard. A giant snail. And, of course, more. Always more.

We are now in a position to track Freud's altered stance towards the animals in his patients' tales. With Emmy, Freud seeks to uncover each animal horror, one by one, and force her to forget it. By forgetting each animal, Freud reasons, she will forget the fear associated with it. In this model, there is a clear separation between the internal world and the external world. Thus, Emmy's fears, which lie within her internal world, become attached to objects – in this case, animals – in the external world. By forgetting the animal-object, her fear must necessarily dissipate since it no longer has an object to attach to. Time and time again, however, Emmy's fear moves to new objects, conjuring up new animal horror stories. Freud's technique, over the course of seven weeks, makes Emmy more functional, but it does not cure her (see also Bromberg 2009; and Tögel 1999). Within seven months, she is ill again.

By the time S.P. consults Freud, his theories and techniques become definitively psychoanalytic. From this perspective, animals are not simply objects in the external world, they are now part of an ambiguous and dynamic inner world of meanings and motivations. The animals are no longer simply 'out there', they are now inside the mind, actively forming its affectual and emotional landscape. In this view, Emmy's reading of the story of the apprentice who feeds a boy a mouse would offer Freud much to ponder. Sure, he would read the story in the paper for himself. But Emmy's active reimagining of the story would make Freud ask her about the mouse, about the mouse being gobbled up, about being frightened to death. He would almost certainly discover fairy tales, childhood fears and anxieties, and perhaps – following Emmy's own associations – even a foreboding 'primal scene'.

In the case of the Wolfman, the 'primal scene' connects, merges and switches the animal and the human in particular ways. On the one hand, this is topographical: the human and the animal are territorialised by separation: human/not human. On the other hand, both the human and the animal undergo processes of transformation, through which the animal and human is reconfigured (as humans become animals and animals become human). This makes sense if animals and humans exist on the same surface and within the same space. This works in both bodily and psychic space. To help understand how physical and psychic space are enmeshed it is helpful to look at Blum and Secor's exploration of psychical and material space (2011).

In their analysis of the relationship between psychical space and material space, Blum and Secor explore Freud's 1909 case study of the Ratman – yet another hybrid form that conjoins the animal and the human. What Blum and Secor discover is how Freud's patient struggles with the idea that his dead father and his lover will be punished for a debt that he owes. Significantly, they argue that the idea of the rat moves through a topological space that connects seemingly contradictory feelings of hostility, love, indebtedness, guilt and love. To account for the ways in which the neurotic Ratman situates people, places and events in the same place, despite their actually being in different places, Blum and Secor utilise the Möbius strip's ability to contain the paradox of having both one and two spaces at the same time. Using this topology analytically, they show how psychical and material spaces shape one another: sometimes by bringing 'things' closer together, sometimes by creating distance between 'things' that are closely related; often, though, by creating a distinction between the inside and the outside.

For my purposes, what is key to the idea of the Möbius strip is not just the paradox of a single surface that always has an opposite side, but also that the Möbius strip can only be fully appreciate by moving along, around and across it. So, let us think about how the Wolfman might exist in a topological space that simultaneously connects and separates animals and humans. On the one hand, the Möbius strip tells us that, at each point, it is clear to the Wolfman (and therefore also Freud) that there is a distinction between animal and human; and, to give it its proper volume, between different animals and different humans. Thus, as the Wolfman travels between ideas about the animal and the human, about animals and humans, he finds one idea leads to the other and back again, leaving the Wolfman (and therefore, also Freud) radically uncertain about what is animal and what is human – even while distinctions between them are often stark. Animals and humans lie on the same surface; they seem to share certain qualities, yet they are never quite in the same place.

The Wolfman, therefore, does not describe a position on the strip, it describes a series of movements, between the animal and the human, that S.P. is continually undertaking as he seeks to give his symptoms, and even his sense of self, a passionate form (following Campbell and Pile 2011). It is the movement along, around and across the surface of animal-human that is significant: it is what allows these supposed oppositions to become syncretic, to merge, to become the ghastly creature known as the Wolfman. It is sheer economy of movement – of affect, of meaning – that makes S.P. become the Wolfman. Yet, the distinction between Wolf and Man always remains: it is not collapsed into a hybrid form, as Freud recognised. So, what Freud seeks, throughout the case study, is to find new passionate forms for S.P., so that he could escape his entrapment in his particular beastly psychical structure.

It is through movement that the Möbius strip reveals itself as a single surface that, topologically, connects seemingly opposing ideas, such as the animal and the human. Yet, at any point on the strip, there remain opposing sides. In this light, Emmy's injunction to 'Keep still'!, and the motionlessness of the wolves in S.P.'s dream, are intriguing because they deliberately arrest movement – but of what? Unconsciously, for both Emmy and S.P., it is by remaining still that they seek to allay their fears of merging with – of being gobbled whole by – the animal. This is not, of course, the only solution to the fear of both being and becoming animal. Strangely, perhaps, another is to imitate them.

Both Emmy and S.P. copy animal behaviour: she, by mimicking the mating call of a bird; he, through his sexual preferences. The protection they seek from being swallowed up by animals is to become beastly, at least in part. Indeed, disguise plays a large role in many of the animal horror stories. As we have seen, animals disguise themselves as other animals (Wolf-goat) and as human beings (Wolf-grandma). Animals are imitative, but so are human beings: Indians dressed as animals horrify Emmy. Along the Möbius strip we travel, never sure what is disguised as what, nor who is imitating what; nor, as Blum and Secor show (2011) 'who is being substituted for *whom*' (p. 1040).

We can say that the human psyche is founded on its entangled, complex engagements with humans and animals, at one and the same time. We can also say that psychic structures in humans are always already founded on precarious and unstable notions of the human and the animal. And we can also say that humans never ever let go of the animal as their psyches develop. Having said all that, what really matters is how and why humans encounter the animal (and other objects) – and draw the animal (and other objects) in as an inescapable and fundamental part of their psyches. *It is not the fact of an entanglement that is important, it is its qualities.* This is what marks the break between Freud's determined attempts to cure Emmy and his rather hopeful interventions to shift the Wolf-Man's pathologies: in Emmy's case, he is not interested in the quality of her relationships with animals and so fails to heed Emmy's injunctions to stop and listen to her; in the Wolf-Man's case, it is precisely the quality of the animal relationships that enables Freud and the Wolfman together to create something called therapy. What animals *do*, Freud realises, is entirely about their shifting place and role in a dynamic and fluid psyche, both conscious and unconscious.

The affective and emotional intensities of human and object relationships must be seen as fundamental to how and why psychical structures take the passionate forms they do. The shifting, mobile intensities of affect and emotion that are evoked by animals (and other objects) are not simply manipulated and modified, nor merely externalised through representations projected onto animals, they are always already part of the beastly mind.

Conclusion: The Twists and Folds in Bodily and Psychic Space

I began with Freud's case study of Emmy von N.: though not his first patient, this is Freud's first case study. At this time, Freud has not yet developed, what he will later call, psychoanalysis. Instead, Freud deploys psychotherapeutic techniques derived from Josef Breuer's supposedly successful treatment of Anna O.; it was Anna O., of course, who famously described this technique as the talking cure. In fact, as we have seen, it is better characterised as kind of fighting-fire-with-fire technique, using hypnosis to systematically stamp out the wildfires of hysterical symptoms. What I drew out from the case study is Emmy's use of animal figures to dramatise her distress. She hallucinates animals; she recalls animal horror stories; she mimics animals; and, she develops a protective formula designed to keep the animals at bay. Keep still! Don't say anything! Don't touch me! I showed that Freud fails to track the animals in his case study, instead focusing on his desire to stamp out the fear and disgust associated with them. The technique, as with Anna O., proves ultimately only partially and temporarily successful. It is its failure, indeed, that prompts Freud to develop psychoanalysis.

In the Wolfman case study, Freud's tactics have changed. Instead of seeking to stamp out S.P.'s associations with animals, Freud tenaciously tracks them, to find out where they go. Freud follows S.P.'s free associations, tracing out his networks of affect and meaning. Instead of discovering some authentic or true original trauma that causes S.P.'s symptoms, what Freud is left with is fantasies, reconstructions, guesses, rationalisations and the like. A 'primal scene' emerges, but only as a kind of mirage that shapes the Wolfman because of the hold it has on him as a vision or thought, not because it is true or real. What is important is that S.P.'s primal scene is populated by both animals and people – by the acts of animals and humans. Indeed, what is animal and what is human can constantly shift, merge and invert, as S.P. searches for appropriate forms through which he can give life to his symptoms. There is no secure ground on which to judge what is human and what is animal, but nor is the Wolfman simply a fusion of the animalhuman.

Observing the fundamental importance of animals in Freud's case studies leads to the question of whether the hybrid formation of the human body-psyche through the animal, expressed even in the naming of the Wolfman or Ratman, without also fundamentally rethinking the human. Significantly, what is at issue is the way hybrids conceal, or deny, both the differences and the movements within the hybrid. On the one hand, we have seen that paying attention to the animals has opened up new possibilities for a spatial understanding of psychical processes and forms. The topological form of the Möbius strip has enabled us to see how the human and animals (and other objects) might be both radical uncertain and also provide certain kinds of ontological truths at one and the same time. As importantly, the Möbius strip attends analysis to both moments of stillness and motionless as well as to movement and dynamism – as the psyche struggles with its beastliness. On the other hand, this confirms that no stable or inevitable demarcation between the human and animals (and other objects) can be held for long; nor can the human be put in a 'black box' that looks like an animal 'black box' (as when people are described as wolves, lions, sheep, foxes, snakes, rats, goats and so on) – or vice versa (as when gorillas, elephants, bulldogs, rabbits, cats and so on are understood to have human characteristics) – without losing sight of the relationality that constitutes and distinguishes the human and the animal.

Taking animals seriously has illuminated why Freud's therapeutic techniques were less and more successful in the cases of Emmy von N. and S.P. It, nonetheless, demonstrates how important objects (such as animals) are for psychoanalytic procedures: where objects are understood not only as human and other-than-human, but also as capable of shifting, merging and inverting the human and the other-than-human. This provides another dynamic for understanding bodily regimes, and their composite corporeal schemas. Bodily regimes are formed out of schemas that 'cut up' and compose (and recompose) the body, however they also cut out certain bodies from others and compose the relationship between them. This chapter has shown how bodily regimes are formed around a human/ other-than-human schema. However, this schema must be understood both topographically and topologically: that is, through processes of transformation and processes that arrest transformation. Moreover, bodily regimes are formed in relationship to passionate forms that provide opportunities to spatialise and demarcate places for desire and fear.

These passionate forms, in part, rely upon social conventions to enable and enact them. We have seen how domestic circumstances and childhood literature help construct a personal aesthetic unconscious for both Emmy von N. and the Wolfman. We have seen how this personal aesthetic unconscious structured their experience and performance of their hysterical, fetishistic and phobic symptoms. However, I said less (than perhaps I should) about the social and spatial context of their lives. Encounters with others occur *in situ*. Indeed, despite wishing to marginalise Oedipus, Oedipus remains a cornerstone for thinking about how the distribution of desire and fear is structured and experienced. In the next chapter, we think a little bit more about this through two very different case studies: Freud's *Dora* and the Hollywood movie *The Hangover*. What links these studies is an Oedipal thread, involving what is now often called toxic masculinity. Following this thread enables me to think more precisely about the extimacy of the unconscious and its role in the distribution of the sensible.

Chapter Five
The Worldliness of Unconscious Processes: The Repressive and Distributive Functions of the Unconscious[1]

Introduction: Following Unconscious Processes, Inside Out and Outside In

It was simply not plausible for Corneille and Voltaire that, after being told by Tiresias that he was the murderer that he was seeking, the penny would not drop for Oedipus (see Chapter 1). How is it possible that Oedipus could not realise, in that moment, that he had killed his father at the crossroads? For Freud, however, the inability to recognise a truth about oneself was entirely familiar, as he was witnessing it in his patients. (And, as we will see, the ability to hide the truth from ourselves is one Freud has in common with his patients. As do we all.) The question, then, becomes how this lack of self-recognition is possible; and, indeed, how it can be so determined as to lead to deafness and blindness in relation to the self. Freud's answer revolves around the idea of a repressed unconscious and the psychic mechanisms of repression, as it is the severity of repression that he says makes his patients sick. Significantly, repression is registered in affects such as disgust, shame, guilt and hate – which, for sick patients, have become unbearable. The repressed unconscious can be understood as a kind of box where traumatic ideas and experiences can be locked away and prevented from escaping, so easing

---

[1] Source: From Kingsbury, Paul, and Steve Pile. *Psychoanalytic geographies*. London New York: Routledge, © 2016, Taylor & Francis.

the unbearability of the affects associated with traumatic ideas and experiences. This explains why, and how, Oedipus cannot see that he killed his father and married his mother. He represses everything that makes his life unbearable. For Freud and Oedipus, this comes at a cost: the punishment is severe, both externally inflicted and self-inflicted.

In this understanding, it is quite easy to see how locked-box repression might align with the assignment of bodies to proper places, with feelings such as guilt, shame, anger and disgust prompting the policing of bodies when they are deemed out of place; or, by circumscribing the distribution of the senses, such that certain aspects of the bodies are considered shameful or disgusting in ways that require policing, such as smell or skin. Repression viewed this way aligns with an aesthetic unconscious of the body that requires and enables policing: thus, we have aligned Freud and Rancière – and this chapter is over before it has begun.

However, we have already seen (in Chapter 4) that repressed material leaks out through bodily symptoms. This tells us that the locked box is not a good metaphor for the repressed unconscious. Rather, we need to be thinking about how repression is achieved as an ongoing process, which is in itself worldly. Therefore, in this chapter, we think further about how repressed material leaks into the world. This aligns, then, with a reworked understanding of the assignment of bodies to proper places and the distribution of the senses, in which more than one bodily regime is in play. Where repression is a set of processes, rather than a set of locked away contents; the assignment of bodies to proper places requires ongoing work to achieve; and the senses get continually worked over to achieve and ensure consistency.

In this sense, what Corneille and Voltaire failed to understand was the sheer unbearability of Oedipus' position: not because it was indeterminate or over-determined, but because it was not; that is, because Oedipus had become trapped. He sought refuge in not knowing. This chapter, then, is about not knowing, as core to understanding how bodies are policed: socially, sensibly and spatially. It is broken into two parts: Freud's Dora case study enables us to think about repression (as the process of not knowing) as produced out of engagement with the world, while *The Hangover* (a Hollywood movie) enables us to see how leaked repressed contents both become distributed through the world and act back on the subject. Let us start by thinking about how intimate unconscious material can be spread through the external world.

In his paper on the 'extimacy of space' (2007), Paul Kingsbury begins by taking on both caricatures of psychoanalysis in geography and also what has been termed the taming of psychoanalysis in geography (by Callard 2003). On the one hand, Kingsbury observes, assumptions have been made about the ways that psychoanalysis is supposed to pathologise entire populations while at the same time deploying a highly personal and decontextualised notion of the subject. On the other hand, he argues that the importation of psychoanalytic concepts seems to have stripped them of the more recalcitrant aspects of subjectivity, such as the difficulty of overcoming traumatic experiences and the ongoing trauma of coping with trauma (see also Blum and Secor 2014). More than this, psychoanalysis has been de-contextualised in ways that appear to absolve society of any blame for those traumatic experiences. In this, the unconscious becomes a particular bone of contention. Its content and structure, formed through, say, the Oedipus Complex, are easily taken to be a general condition, while at the same time material in the unconscious (such as heather cocks and wolves) can be uniquely personal. Kingsbury's solution is to turn to Lacan's topological understanding of space to understand the relationship between inside and outside (as do Blum and Secor 2011).

In Chapter 4, we saw how the inside-and-outside look from the perspective of two individuals, Emmy von N. and the Wolfman. In this chapter, we explore the ways that the outside world generates unconscious material and the way that unconscious material is spread through the world. This adds volume to our account of the Möbius strip, as it become more social, more extimate. Thus, Kingsbury introduces two ideas that are critical for this chapter. The first idea relates to the flow of desire and fear from inside out and from outside in:

> On the one hand, [there is] the transference of people's intimate feelings, thoughts and beliefs on to an external object, and, on the other, [there is] the stirring and blooming of people's inner feelings, thought and beliefs by an external object. (Kingsbury 2007, p. 245)

As we know, this flow inside out and outside in is given conceptual form by the Möbius strip (and other topological figures, such as the Klein bottle and torus). Lacan's neologism extimacy is specifically intended to capture the ways that people's intimacies become projected onto, or played out through, the external world, but also how the external world structures and prompts people's interior worlds. We saw this in operation in Chapter 4. In this chapter, we focus more upon the second idea that Kingsbury emphasises. For him, significantly, the term *extimacy* also bears a relationship to the uncanny. Central to this is the way that the world can conjure up or prompt unconscious ideas for the individual; this conjuration makes the world strange. Thus, in uncanny experiences, the familiar is made strange.

However, the uncanny – and therefore extimacy – implies more than this (see also Kingsbury 2017). In uncanny experiences, the familiar is rendered strange in ways that make people feel uncomfortable, disturbed, disorientated, unsettled, creeped out. It is registered in the body (as Straughan has shown, 2014). It raises the hairs on the back of your neck, it creates goosebumps (but not in a good way), hands grip the mouth, the body instinctively tightens. The uncanny is a species of fear. Thus, extimacy traffics affects through interior and exterior worlds, but in exteriorising internal worlds they become strange and unfamiliar. This means that, for Kingsbury, 'our most intimate feelings can be extremely strange and Other to us' and that 'our feelings can be radically externalised on to objects without losing their sincerity and intensity' (p. 235).

This creates a model of engagement with the world in which the psychical and the social loop into and out of one another (Bingley 2003). However, the idea that we can become alienated from our own thoughts and feelings as well as experience our own thoughts and feelings as if they were strange to us interferes with the linearity and connectivity of the Möbius strip. The whorls of affects around and through interior and exterior worlds are not necessarily linear or free-flowing, nor do they happen just as people please. There are structured. And they can be traumatic. Creating cuts, bypasses and dead ends in the circulation of affects. So, along with idea of the movement of desire and fear around the Mobius strip of inside and outside, we must also develop a model of the unconscious that is both open to these flows, but also closed off or semi-permeable or diverted. For this, the model of the repressed unconscious is useful, as it captures those moments in which unconscious ideas are captured and contained. This means paying attention to Freud's repeated observation that he was primarily interested in the repressive unconscious because it was repression – and repressed material – that made his patients sick (Freud 1915; see Campbell and Pile 2010).

Significantly, repression makes Freud's patients sick because it is worldly, not just because it has specific harmful psychic contents. Bluntly, the unbearability of traumatic ideas and experiences – and its repression – is both personal and social. To understand the worldliness of repression, I explore two aspects of Freud's encounter with Dora. First, I set it in the context of the history of hysteria (the umbrella term used to describe symptoms in the late nineteenth century). Second, I look at Freud's encounter with Dora. In particular, I am interested in how Dora made Freud think about the role of the repressed unconscious in therapy, as this illuminates how repression is entangled with the world.

In this chapter, the touchstone for these entanglements is Oedipus: in part, because Oedipal relationships are a product and productive of patriarchal rela-tionships; in part, and related, because one of the hallmarks of the Oedipus myth is castration and castration anxiety. In Dora's case, her entanglements were dom-inated by what we might call patriarchy, especially in the form of the exchange of women amongst men, including Freud. Patriarchy both imposes traumatic expe-riences upon her and forces her to repress those experiences. While Freud became increasingly aware of Dora's repression and the expression of repressed material, he remained deaf and blind to his own role in her patriarchal relationships. Meanwhile, *The Hangover* hides and exhibits its castration anxieties, almost as if Tiresias had written the script. Both Dora and *The Hangover*, then, are Oedipal stories, connected to repression, experiencing and performing their traumas and anxieties in the world, but unable to hear or see them.

Where Dora's story is focused on her engagement with, and attempts to nego-tiate, a repressive external world, the movie *The Hangover* is predicated on the idea that there is a place in the United States where repression is foreclosed. That place is Las Vegas. In Las Vegas, anything goes. Whatever is done there, stays there. Las Vegas becomes a spatialisation of the possibility that there is an unrepressed place that can act as the playground for taboo ideas freed from the unconscious. This is a proposition. It sets up a question: if there is a place where there is no repression, then how would we understand processes of repression in such places?

The Hangover, further, affords an opportunity for thinking spatially about the unconscious: that is, for showing that the unconscious always is disturbing the boundary between the subject and the external world, simultaneously organising that boundary and subverting it. This underscores a notion of the unconscious that is both social, in the world, and also deeply personal to the subject: yet where both are extimate (as Paul Kingsbury puts it, 2007) and also distributed spatially (in this case, through Las Vegas). Put another way, *The Hangover* gives me an opportunity to think about, on the one hand, the spatial production of the unconscious and, on the other, the unconscious production of space.

Taken together, the case studies map out the distribution of the unconscious. Here, I am explicitly echoing Rancière's notion of a distribution of the senses. However, distribution in this chapter is an attempt to grasp the territorialisation of the unconscious, rather than deploying Rancière's more structural sense of distribution. Or, put another way, it is to emphasise that the idea of distribution as inherently spatial. That is, the unconscious – its dynamics and content – can only be understood contextually. Even the ways that social becomes intimate and the intimate becomes social are set in social contexts that determine how people can receive or express unconscious ideas. This is important, for this is about creating a model of the unconscious that enables an understanding the transfer of affects between people – an idea that we pick up in Chapter 6.

In this chapter and the next, then, I am building a model of the subject that sees it as radically open to the world – even with all its closures and borders. This openness is radical because it structures even the most personal parts of the unconscious, at the very root of subjectivity. Thinking through this requires, in the first instance, a model of the unconscious as the product of a set of processes, which, we will see, are social, spatial and personal.

Hysterical Symptoms and the Unconscious: Dora, What She Does Not Know or Cannot Tell Us

The idea of the repressed unconscious – and its significance for understanding what makes people sick – has its origins, not in a pre-existing conceptual framework or a pre-determined model of human psychical development, but in the external world. This is as true of Freud's patients' symptoms as the psychoanalytic concepts and techniques that have emerged over more than a century to understand them. Let me be clear, the idea of a repressed unconscious emerges in response to some extremely puzzling medical problems that had already baffled medicine for well over 150 years before Freud. In particular, hysterical symptoms, where people's distress was expressed through the body in often bizarre and seemingly theatrical ways, were particularly puzzling as they could not be directly correlated with brain disease or other physical characteristics. In 1885, Freud is in Paris, with Jean-Martin Charcot, seeking to understand whether abnormalities in different regions of the brain had any connection with hysteria (see de Marneffe 1991; and Sulloway 1979, pp. 28–30). What Freud witnesses is astonishing: see Figure 5.1.

What we see in this painting is Charcot proving – to a fascinated audience – that the origins of some nervous disorders, such as hysteria, do not lie in the body, but solely in the mind. He is demonstrating that a patient's symptoms could be turned on and off, literally, by the gentle touch of a hand. Put another way, what clinicians were (and still are) faced with was a wide variety of physical symptoms that appeared to have no foundation in physical problems, despite a prolonged search for explanatory physical abnormalities in autopsies of dead patients (with particular emphasis on the search for lesions in the brain – which was Freud's own particular interest, too, although he cautioned against a too localised understanding of brain functioning). You will note the instruments that lie on table by the standing central figure, Charcot. These consisted of probes and electrical stimulation devices. Using these, Charcot demonstrated that his patients' symptoms

Figure 5.1   *A Clinical Lesson by Doctor Charcot* (1887) by Pierre-André Brouillet. Source: Musée d'Histoire de la Médecine, Paris.

were not simply faked or simulated. Paralysed limbs really were paralysed, beyond the conscious control of the patients. Yet, using hypnosis, Charcot could make the paralysis completely disappear (and reappear). Here was proof of the power of the mind to dominate the body – not consciously, but *unconsciously*. Returning to Vienna, brimming with enthusiasm and confidence, Freud sets about developing a cure for hysteria using hypnosis.

In collaboration with Joseph Breuer, Freud imposed therapeutic regimes consisting of daily massages, rest and relaxation, restricted diets and hypnotic sessions. Despite the mixed results, Breuer and Freud presented case studies that indicated a confidence in hypnosis as the cornerstone technique for a complete cure for hysteria (Freud and Breuer 1985). Indeed, it is Breuer's patient, Anna O., who coins the expression 'talking cure' so often (yet wrongly) associated with psychoanalysis. Yet, Freud remains doubtful of the success of hypnosis. For example, the curative effects appear to wear off, lasting, at best, a few months. Within a year, some of his patients were back in therapy (notably Emmy von N.) with different doctors. Embarking on a radical and punishing period of self-analysis in the wake of the death of his father in 1896, Freud becomes increasingly confident of the benefits of approaching his patients' symptoms through an entirely different technique: dream analysis. Indeed, it is in 1896 – some 11 years after his encounter with Charcot's hysterics – that Freud begins to call his therapy psychoanalysis. Three years later, late in 1899, a young woman has been brought to Freud by her father. She is suffering from a variety of symptoms, which appear not to have physical origins. In his case history (misleadingly, published in 1905), Freud will call her Dora – who we now know to be Ida Bauer, 1882–1945 (see, especially Mahony 1996).

In October 1899, Freud began treating a 17-year-old woman, Dora, who was suffering from a variety of symptoms. (Dora turned 18 on 1 November 1899.) As it happens, Dora lived on the same street (at 32 Bergasse) as Freud, so he already had already witnessed her coughing and hoarseness in the summer of 1897. Yet, she had also suffered from a feverish condition in the winter of that year, which was initially (wrongly) diagnosed as appendicitis. By 1898, Dora, according to her parents, was suffering from mood swings and character changes. As Freud puts it:

> she was clearly no longer happy either with herself or with her family, she was unfriendly towards her father and could no longer bear the company of her mother, who constantly tried to involve her in the housework. She tried to avoid contact with anyone; in so far as the fatigue and lack of concentration of which she complained allowed, she kept herself busy by attending public lectures, and devoted herself seriously to her studies. (Freud 1905, p. 17)

Freud presents Dora as an intelligent woman, who is curiously beset by contradictory motivations. Typical teenager, you might think. Except, what really shocked Dora's parents was a suicide note they discovered. While they did not think that Dora was serious in her intention to commit suicide, they were nonetheless horrified. On top of this, she was suffering from blackouts and amnesia. For Freud, Dora was an 'ordinary' case, exhibiting common hysterical symptoms, such as 'dyspnoea [shortness of breath], tussis nervosa [nervous cough], aphonia [loss of voice], along with migraines, mood swings, hysterical irascibility and a tedium vitae [weariness of life] that is probably not to be taken seriously' (1905, p. 18). We can almost hear Freud yawn, but what piques Freud's interest is the possibility of using Dora as an addendum to his *The Interpretation of Dreams* (1900). She is to be the proof of his dream pudding. Dora, however, is not going to play along. It is in her resistance to Freud that we can begin to see how repression and hysterical symptoms provide Dora with a response to, and a diagnosis of, the patriarchal relationships that envelope her.

Much has been written on the Dora case study, including a vast array of academic studies (Cixous 1976; Moi 1981; Roof 1989; Bernheimer and Kahane 1990; Decker 1991) as well as stage plays, novels and art books (Morrissey 1995; Yuknavitc 2012; and Kivland 1999, respectively). There is little sympathy for Freud in most of these engagements with the Dora case study, seeing him (basically) as an agent of patriarchy. My purpose is not to save Freud from his critics, as they all have a point. Rather, I wish to illuminate the emergence of Freud's concept of the unconscious in the course of his analysis of Dora. It will take 15 years, after Dora, for Freud to formulate his concept of the unconscious, but in this case study, we can already glimpse its key features – importantly, as a response to the experiences and symptoms that Dora suffers.

Though Freud's analysis turns upon an analysis of two dreams, I will use the incident by the lake as the focal point of our analysis. This gives us a stronger grip on the extimacy of the unconscious. More than this, it also shows that, while Freud is increasingly focused on what we might call the repressed unconscious (the form of the unconscious that is most closely identified with psychoanalytic thought), there is more than one kind of unconscious in play. In Dora's dreams and experiences, what we are listening for is those moments when the truth is being hidden or disguised or misrecognised. These moments attest to repression in operation. In her symptoms, moreover, we are witnessing unconscious processes redistribute Dora's distress through her body. As you might expect, the traumatic experiences and ideas are associated with sex.

So, what happened by the lake? And what might this have to do with Dora's coughing and breathing problems? When Dora was 16, she went on holiday with her father, who she calls Papa (real name, Philip Bauer). After a few days, Dora's father intended to return to Vienna. The plan was for Dora to spend several weeks with Herr and Frau K. (who we now know to be Johann and Peppina Zellenka).

> But when her father prepared to set off, the girl suddenly announced very resolutely that she was going with him, and she had done just that. It was only some days later

that she gave an explanation for her curious behaviour, asking her mother to inform her father that while they were walking to the lake to take a boat trip, Herr K. had been so bold as to make a declaration of love to her. (Freud 1905, p. 19)

After being informed of her explanation, Herr K. and Papa's response to Dora's mother (Käthe Bauer) is predictable: denial and refusal. Herr K. denies that he had done anything that would even have permitted Dora such an interpretation of his behaviour. For his part, Papa declares that he believes the whole incident to be a fantasy. Indeed, Frau K. is blamed for the fantasy: she had talked too deeply and intimately with Dora about 'sexual matters', and even allowed her to read Paolo Mantegazza's *Physiology of Love* (published in 1896), which, according to Papa and Herr K., had clearly inflamed Dora's erotic imagination.

The erotic entanglement between Dora and Frau K. is, in fact, complicated by Papa, who also shares an intimate relationship with Frau K. So, when Dora demands that her Papa sever his links to both Herr and Frau K., he steadfastly refuses to do so. In particular, he defends his relationship to Frau K., which he says is 'an honest friendship' that can do nothing to hurt Dora (1905, p. 19). Frau K., he says, is suffering very badly from her nerves and he is her sole support. Indeed, as he 'gets nothing' from his wife (Dora's mother), his friendship with Frau K. is also a comfort to him. The picture presented to Freud, then, is of two entirely innocent men, Papa and Herr K., neither of whom have done anything to harm Dora. Dora, it seems, is a victim of her own overly sexualised imagination, caused by the overly intense and intimate nature of her relationship with Frau K.

Freud is instantly suspicious: there must, he reasons, be more to the story. Indeed, there is. Freud solicits from Dora an earlier experience with Herr K. when she was just 14.

After the first difficulties of the cure had been overcome, Dora told me of an earlier experience with Herr K., which was even more apt to act as a sexual trauma. She was fourteen years old at the time. Herr K. had arranged with Dora and his wife that the ladies should come to his shop in the main square of B. [Merano, northern Italy] to watch a religious ceremony from the building. But he persuaded his wife to stay at home, dismissed his assistant and was on his own when the girl entered the shop. As the time of the procession approached he asked the girl to wait for him by the door which opened on to the staircase leading to the upper floor, as he lowered the awning. He then came back, and instead of walking through the open door, he suddenly pulled the girl to him and pressed a kiss upon her lips. [. . .] But at that moment Dora felt a violent revulsion, pulled away and dashed past him to the stairs and from there to the front door. After this, contact with Herr K. none the less continued; neither of them ever mentioned this little scene, and Dora claims to have kept it secret even at confession at the spa. After that, incidentally, she avoided any opportunity to be alone with Herr K. (Freud 1905, p. 21)

Freud's response is alarmingly patriarchal: instead of understanding Dora's reaction as perfectly normal response to an unwanted and forced sexual advance, he wonders why she is repulsed and runs away. Even so, what emerges later in the case study is a far more complex picture of the love and sexual relationships in Dora's world. The problem, for Dora, is that she is secretly in love with Herr K. Part of her motivation, then, in wishing Papa to break off his affair with Frau K. is so that Herr K and Frau K can divorce. Yet, Dora is also in love with Frau K., so she does not want Papa to simply replace Herr K. in her life. To solve the problem of loving both Herr and Frau K., Dora wishes that Papa would return home and she rejects Herr K. Without saying so, she chooses Frau K.

It is not that Dora is confused by all of this: the opposite, in fact. Dora's problem is that she is too aware of these relationships – and this forces her to repress (as an unknowing process of unknowing) her feelings and thoughts – and, to this end, she replaces one idea, a kiss, with another idea, a cough. Her bodily symptoms of coughing and difficulty in breathing, which started at the time of the lake incident, refer back to her repulsion of the kiss. More than this, the coughing is also connected to Dora's witnessing of how her Papa was treated when he was ill with a severe cough. This is no innocent illness, for Dora, as she assumes that it is connected with her Papa's syphilis (which Philip contracted prior to marrying Käthe). Worse, Dora assumes that her Papa's syphilis has been passed on to her (1905, p. 65). In the analysis, then, Freud is constantly trying to follow the twists and turns in Dora's sexual, erotic and emotional relationships – many of which are hidden by secrecy or forgotten through repression.

Alongside the twists and turns of Dora's sexual worlds (involving her own emerging sexuality, Papa, her mother, a governess, Herr K. and Frau K.), Freud notes various ways in which unconscious processes work to hide or repress the impossible demands of these worlds as Dora experiences them. One unconscious process concerns 'contiguity' and 'the temporal proximity of ideas' (p. 30). According to this, ideas are associated with one another through some kind of closeness. Thus, through her cough, Dora demonstrates not only her imitation of her Papa, but also her love of Herr K. – her attacks coincided with times when Herr K. was absent on business (p. 30). Another feature of contiguity is the closeness between the repressed idea and the physical symptom: love for Papa and love for Herr K. manifests itself as a cough and breathlessness because there is a direct link between Papa and Herr K. and the symptom – syphilitic illness and the repulsive kiss. There are, further, contiguities between Papa and Herr K.: not just through Frau K., but through Dora herself. Dora refuses, angrily, to be 'the gift' that Papa gives Herr K. to buy his acquiescence over the affair with Frau K. The cough and breathlessness are, as with all symptoms, *overdetermined*: that is, determined by more than one idea. Freud says, 'a symptom has more than one meaning, and serves to represent several unconscious trains of thought' (p. 36). Indeed, it is this multiplicity of thoughts – and their movement along chains of association – that creates enough 'force' to create a symptom. So, another key aspect of unconscious processes is the way that thoughts move and interact, dynamically, with one another.

Further, it is this coexistence of contradictory ideas that creates the kind of intensity that requires repression. Thus, if two ideas coexist excessively intensely – I want Papa to love me more than Frau K.; Papa's sexuality frightens and disgusts me – this itself can cause repression, such that both ideas become unconscious. These contradictory unconscious thoughts, because of the intense affects associated with them, gain enough force to create symptoms: only, dynamically so. Thus, the symptom can readily take on new meanings. Similarly, the same thoughts can create new symptoms. Moreover, the unconscious accommodates and preserves both thoughts and experiences from different times of life at the same time and also contradictory thoughts and ideas. Thus, for Dora, 'I love Herr K.' and 'Herr K. disgusts me' exist side-by-side without conflict (p. 43). The coexistence of contradictions, perhaps drawn from different traumatic ideas and experiences, means repressive processes do not require consistency or logic to be effective.

A significant feature of unconscious processes is the conversion of one idea into its opposite. Freud uses the metaphor of a pair of astatic needles (yet another spatial metaphor) to describe how conscious and unconscious thoughts can run in exactly the opposite direction, yet be driven by the same force and point along the same line as one another (p. 43). The significance of unconscious processes that can contain contradictory ideas and reverse ideas into their opposites is difficult to overestimate. For, these can manifest themselves in very ordinary symptoms such as confusion, ambivalence, internal conflict and aggressivity – as with Dora's mood swings and apparent changes in character. Importantly, these mood swings are consistent in Dora's internal world, the problems arise from their interaction with her external worlds – it is not just that her parents do not understand her, or that social conventions have forced her to hide or repress aspects of her sexuality (such as her masturbation and, perhaps, her lesbianism); on top of all this, she is struggling to handle the all-too-real sexual experiences and desires that she encounters in the world. Her consistent response is, simply, to deny that she knows what is going on (p. 46) because knowing what is going on is traumatic.

The unconscious, then, is not a place – as if it were a locked box inside the mind where we put all the painful stuff; nor is it a set of contents – as if the locked box only contained repressed sexual desires associated with the Father or Mother. Importantly, we have seen that there are two distinct functions creating unconscious contents. Of course, there is repression of affects-and-ideas (which is generative of Dora's symptoms), yet alongside this there is also a more conscious function, secrecy (which is no less harmful in its effects). Rather, the unconscious is best associated with processes: such as the contiguity of ideas, reversal into opposites, repression, overdetermination, trains of thought, chains of association, timelessness, contradictoriness and preservation. Yet, the unconscious, for Freud, is not simply a bundle of processes. There are different kinds of unconscious dynamics, each characteristically with its own spatial organisation within the psyche.

Freud's Topographical, Dynamic and Economic Understandings of Psychical Processes

Freud's first fully developed account of the unconscious appears in 1915 (although his first attempt to outline the concept of the unconscious was published in 1912; see Campbell and Pile 2010). *The Unconscious* is best known for spelling out three different aspects of the unconscious. To begin with, Freud argues, it is important to consider the psyche as having a *topography*. Psychic topography has three different systems: the conscious, the preconscious and the unconscious. For Freud, it is important to understand not only in which system a psychical act takes place, but also how psychical acts move, or do not move, between systems. His concern is not only with psychical material, but with its psychic location and with its mobility or lack of mobility in the topography of the mind (see 1915, pp. 56–57). To understand the forces that move psychical acts, or prevent their movement, within the psyche, Freud argues that it is necessary to supplement the topography of the psyche with a *dynamic* understanding of psychical processes.

To explain the dynamism of psychical processes, Freud adds a theory of drives to his model of the psyche. Famously, these drives are bundled around two opposing (yet related) poles. For our purposes, it is helpful to think of these bundles, crudely, as *sex* and *aggression*. Psychodynamics, from a Freudian perspective, involves both the affectual forces associated with the drives within the psyche, and also the interaction between affectual and emotional forces within and between systems. To account for the ways these forces work within the psyche and between systems, Freud adds an *economic* perspective. Using this idea, Freud seeks to understand how the psyche makes investments in particular ideas and gains a return from that investment. In other words, he tries to discover the fate of 'quantities of excitation' within the psyche (1915, p. 64). Freud's main concern, however, is not with 'quantities of excitation' in general, but with a particular excitation: that is, with what he calls *libido*, or the sexual investment associated with the sexual drives. What is key is whether the fate of the sexual drive is to be repressed or not. In repression, 'the idea representing a drive [is] not removed or destroyed, but prevented from becoming conscious' (1915, p. 49). Bluntly, it is both the persistence of the idea that re-presents the (sexual) drive *and* its repression that causes his patients to become so sick. In the case of a child with an animal phobia, this will cause anxiety 'in two situations, first when the repressed love impulse becomes intensified, second when he perceives the feared animal' (Freud 1915, p. 65).

Freud's example of the phobic child is almost certainly the Wolfman (Freud 1918; see also Chapter 4). In this case, simply put, the 4-year-old Wolfman represses how his sexual drives have become entangled with his mother and father by hiding an idea, about sex between his parents, behind an even more intense image, that of frightening wolves – as wolves constantly threaten to gobble up little children. So, as an adult, the Wolfman experiences neurotic anxiety not only in relation to sex, but also in relation to animals. Thus, unable to deal with the fate of his sexual drives, the Wolfman represses his sexual affects and emotions – but in such a way that causes him to become sick.

Yet, Freud's analysis opens up the unconscious to the Wolfman's world (see Chapter 4). Thus, what makes these sexual drives so painful and unacceptable for the Wolfman is determined by his relationships both with his family (including his sister) and also other adults (such as his nurse). More than this, the fate of the Wolfman's drives is also determined by his relationship to his animal world (not only wolves). What Freud opens up is an account of the unconscious that is more than psychical, more even than familial and social, it is also distributed through relationships to other things *and* through the Wolfman's worlds. Alongside family and social mores, alongside his internal psychodynamics and animal phobias, the Wolfman's neuroses are also formed in relation to his wealthy family's country estate and large house (see Campbell and Pile 2011).

For Freud, there are highly differentiated unconscious processes. His topographical understanding of the mind creates regions in the psyche: conscious, preconscious and unconscious. This topographical understanding will be supplemented by a further territorialisation of the mind into regions: super-ego, ego and id. What is significant is that these two territorialisations of the psyche exists side by side, yet do not simply or fully map onto one another. Arguably, they do not 'add up' into a coherent, integrated and consistent cartography of the psyche. Just as Freud charted out different processes for the body (e.g. skin-ego) and mind (e.g. the pleasure principle), so we must also be clear that the unconscious is not a stable, coherent, integrated place within the mind. Instead, we must think about how it is produced by multiple determinations (one of which, to be sure, is repression) many times over.

To repeat, the unconscious is comprised of more than just repressed ideas. Freud is clear about this: 'the repressed does not constitute the whole of the unconscious. The unconscious is the more extensive; the repressed is one part of the unconscious' (1915, p. 49). It is by extending our understanding of the unconscious beyond repression, that we can more clearly see the distributive functions of the unconscious, both producing and produced by its entanglements with the external world. Viewed simply, if the unconscious was only about repression, then material stored there would never escape. The forces of repression would ensure no troubling affects would ever leak into the body, into the mind, or into the external world. However, this is not the case. And, since it is not the case, the unconscious must engage with the external world. I argue that this entanglement is structured spatially by its distributive function.

Dora has told us how repressed affects continually find their way into people's bodies, thoughts and lives – albeit in disguised ways. In other words, repressed ideas are in circulation and are worked on by other unconscious processes, including disguising them and hiding them behind other ideas. Repression does not take place away from the world, only in people's heads, but in its engagements with it, experienced through people's bodies. In this light, the distribution of the senses is structured by processes of repression that determine where the boundaries between knowing and not knowing, feeling and not feeling lie, where these boundaries are drawn in psychic space, but also social and bodily space.

Dora shows us that repressive processes are operationalised through the body and in engagements with the world. Her symptoms show that repressed materials are not contained, but find alternative ways to express themselves in the world. This takes us back to Kingsbury's notion of the extimacy of space: the leaking of intimate ideas into the external world. To explore this leaking, I use the fantasies played out in the film *The Hangover* (2009). This movie is useful partly because the hangover is a metaphor for forgetting (both hidden through secrecy and suppressed through repression) and partly because it takes place in a supposedly unrepressed place, Las Vegas. This setting is highly significant (as is Bangkok in the sequel). Las Vegas is a city long associated with its supposed lack of repression of people's desires: yet, what happens in Vegas is meant also to stay there, as if Vegas functioned as the location for America's repressed unconscious. It is important that repressed ideas (around masculinity, sex and marriage) find their way into the world through this film, too.

What Happens in Vegas? When the Fates of Drives Become (All Too) Real

The plot of *The Hangover* centres on a bachelor party gone badly wrong. The premise for the film is quickly established. The friends have five hours to find the groom, Doug, who is lost, and get him to the wedding on time. Will they make it? First, of course, they must find Doug. It is in the unfolding search for Doug that the comedy – which is mainly situational and slapstick, but also the product of the stupidity and crudity of the characters, rather than them being clever or witty — transpires. The movie is funny, but I had thought no more about it until Virginia Blum perceptively observed, over pizza and merlot, that the film was like a psychoanalytic session. I take this insight to be an open invitation to reconsider the film from a psychoanalytic perspective, but I also want to think about it spatially. First, a plot recap (spoiler alert!).

Doug Billings is marrying Tracy Garner. Doug is being taken to Las Vegas for one last bachelor weekend. His best friends are Phil Wenneck, a high school teacher, and Stu Price, a dentist. Doug persuades Phil and Stu to let Tracy's brother, Alan, accompany them. Before they set out, Sid (Tracy's father) and Doug talk about the trip. Sid reminds Doug that 'what happens in Vegas stays in Vegas'. Doug laughs, uncomfortably. Then, with a face that suggests bitter personal experience, Sid adds, 'except for herpes – that shit will come back with you'. The young men drive to Vegas in Sid's beloved Mercedes. They arrive in Vegas just as day turns to night, and their desire to take advantage of Vegas' many delights increases as they take in the bright neon lights of the city's hotels. At Caesars Palace (a hotel, not a Roman Palace), Phil decides that they should spend the night in a $4,200-a-night villa (on Stu's credit card). The view of Vegas at night is intoxicating (Figure 5.2). On the rooftop of Caesars Palace, the men toast their friendship and the night ahead.

The next morning, Stu, Alan and Phil painfully stir from unconsciousness to find their expensive suite in chaos. Amongst the wreckage of the night before, the hung-over men discover various empty bottles and cans, a smashed TV, a smouldering chair, underwear, a sex doll and a blow up pig in the Jacuzzi, a chicken. . .a

Figure 5.2   Las Vegas at night, 2009. Source: Steve Pile.

baby. . .and, as Alan goes to pee, a tiger. Stu is missing a tooth. Worst, Doug is nowhere to be found. And they cannot remember what happened in Vegas.

Las Vegas, of course, is the setting proper for the bachelor party. The image of Vegas as a 'sin city' has been carefully nurtured since the 1960s, built on the twin pillars of sex and gambling. The city seems free – offering 'free' pleasures and 'free' money, spiced with liberal amounts of alcohol and other drugs (including testosterone). The often-repeated expression 'what happens in Vegas stays in Vegas' seemingly lifts the restrictions of normal moral behaviour, yet also puts a firm (social, spatial and temporal) boundary around this amorality. In the city, anything goes, but outside the city whatever happened must remain, if not forgotten, then at least unspoken. Yet, as Sid's warning about herpes demonstrates, what happens in Vegas must leave no trace – such as marriage certificates, tattoos, photographs, sunburn and missing teeth – otherwise what happens will come home with you. Vegas is paradoxical, then: offering the freedom to act out your fantasies, but only if you follow the rules set by the inevitable return to real life. Perhaps unsurprisingly, it is this tension, or contradiction, between 'reality' and 'fantasy', so intense and visible in Vegas, that has so caught the attention of architectural and urban theorists (Venturi, Brown and Izenour 1972; Hannigan 1998). Indeed, what happens (before and after the bachelor party) to Phil, Alan and especially Stu traverses the fine line between the fantasy that sin city is fun, sexy and consequence-free and its somewhat scarier, repercussive reality.

In the pool area of the Caesars hotel, Phil, Alan and Stu make their first efforts to recall the night before. In a scene where Alan makes the baby simulate masturbation (twice), various tanned fat-free young women in skimpy bikinis are arbitrarily shown, and Stu vomits at Phil's feet, normal social conventions are crudely transgressed, without apparent humour. Amongst all this, the trio attempt to reconstruct the sequence of events. They remember the rooftop toast, the dinner at The Palm, then craps at the Hard Rock. That's the last they recall of the night, and Doug. Now, they have to deal with not knowing. At a loss, they check their pockets, discovering an $800 ATM receipt at the Bellagio with the time stamped at 23.05 and a valet ticket for Caesars clocked at 05.15. On Phil's wrist, there's a hospital tag. Less of a chronology, the men now have a preliminary map of the night before. Their first stop is the hospital, where they discover that their inability to remember is due to taking 'roofalin' (the movie's slang for rohypnol). Rohypnol has gained notoriety as the date-rape drug, so even though the hospital confirms that Phil has not been physically raped, both Phil and Stu nonetheless seem to feel psychologically raped.

The hospital doctor gives the trio their next clue, apparently one of the group got married at the Best Little Chapel. It turns out that it is Stu who got married. So, they leave to find his new wife, Jade, who is also the mother of the baby. As they drive away, their car, a stolen police car, is attacked by two men, wielding an iron bar and a baseball bat. 'Where is he?' one of the attackers yells. Phil hits the accelerator and speeds off.

Clueless and dejected, they decide to confess to Tracy that they have lost Doug, but before they can call they are sideswiped by a large black SUV. A small Asian man, Mr Chow, approaches their wrecked car, accompanied by two henchmen. He demands that Phil, Stu and Alan return his money, $80,000 dollars, in cash, or he will kill their friend, who he has captive. On a promise that Alan can 'beat the system', the men head for the Riviera casino to win the money at the blackjack tables. To everyone's amazement, it turns out that Alan is a blackjack savant. Quickly, they win $82,400 before the casino cottons on and they make a sharp exit.

With real hope in their hearts, they head for their desert rendezvous with Mr Chow. The men exchange the money for Doug, but when they take off the hood they find, to their horror and dismay, that it is not their Doug, but another Doug – in fact, the drug dealing Doug that sold the bad drugs to Alan. It is only then that Phil calls Tracy to tell her that they have lost Doug, and that the wedding is not going to happen. It is then that the actual clues to Doug's whereabouts finally fall into place. Like a good dream analyst, Stu connects the dots differently. Thanks to something the wrong Doug says, roofies makes Stu think of ending up on the roof. Stu realises where the right Doug is: on the roof of Caesars Palace.

Finally, they rescue a thoroughly sunburnt, and furious Doug. They have three and a half hours to drive to the wedding. Of course, they make it, just in time. Only Doug's red skin suggests anything is awry and that their perfect wedding is not exactly as it should be. After the wedding, Alan shows the other three men photos taken on Stu's camera, showing them exactly what did happen in Vegas. The four agree to look at the photographs *one time* only, afterwards they will 'delete the evidence'. The last picture on the camera shows the four men, in blue wedding suits, at the Best Little Chapel.

So far, we have seen Phil, Stu and Alan follow clues in an attempt to find the secret to Doug's whereabouts. In some ways, this can be seen as analogous to the interpretation of a dream. The various elements of the dream act as clues, which lead to further clues. As the various webs of meaning are traced out, so the riddle to the dream's core concern can be discovered, but only with forensic and painstaking attention to detail. That said, the clues in *The Hangover* also trace a psychogeography of Las Vegas, entwining desire and fear, sex and money, lost and found. Las Vegas, in some ways, becomes a map of the unconscious. This is not simply, or only, a repressed unconscious, however. Las Vegas contains elements that are overlooked (such as the relationship between the roof and the roofies) as well as destroyed and forgotten (rendered unavailable to memory by the roofies – where did the chicken come from?).

The night seems to have been wild and unrepressed, taking in luxury hotel rooms, drugs, alcohol, a strip club, a stolen police car, gun fire, gambling, prostitution, encounters with celebrities, a pet tiger and a sex act in a hotel lift. These are all unconscious, but none are truly repressed. Rather, they are forgotten or overlooked or tantalisingly just out of reach or on the tip of the tongue or require considerable effort to be recovered. Yet, it is through these seemingly random and disconnected elements that the repressed unconscious can find passionate forms to express itself (not unlike the process of dreaming). If nothing seems to be repressed in Vegas, then this is only because the repressed unconscious finds ample opportunities to find forms for its passions via the distributive functions of the unconscious. To get at this, I focus on marriage in the film, for this provides sufficient opportunities for unconscious ideas to break cover and express themselves in the world.

The Distribution of Unconscious Material: A Tale of Two Weddings, Three Women and Four Men

Tracy and Doug are, evidently, the perfect couple, ready and willing to turn their perfect wedding into a perfect marriage. Sure, Tracy is 'freaking out', but her anger and anxiety are entirely appropriate under the circumstances. Tracy stands in marked contrast to Stu's girlfriend Melissa. When Doug, Phil and Alan pick up Stu, Melissa is crossly barking orders at Stu. He has to take his Rogaine®, and not forget to use it. He must call as soon as he gets to the hotel. She does not want to be kept waiting. Her voice is sharp and mean-spirited. Old misdemeanours are brought up as Melissa ensures that Stu knows she will be monitoring him. Stu sighs and complies. Melissa doesn't want Stu going to some strip club. Stu reassures Melissa that they are going on a harmless trip to wine country. There will be no bachelor party behaviour – and definitely no strippers. Melissa despairs: 'It's just boys and their bachelor parties, it's gross. . .not to mention it's pathetic, those places are filthy. . .and the worst part is that little girl grinding and dry humping the fucking stage up there, that's somebody's daughter up there'. Stu parrots Melissa: 'that's somebody's daughter up there'.

During the drive to Vegas, we learn about Phil's life. He left his wife and kid behind. And he is really happy to do so. Phil hates his life. He'd rather stay in Vegas than go back. Phil turns to Doug: 'You know what, Doug, you should enjoy yourself. Come Sunday, you're gonna start dying. Just a little bit. Every day'. Nonetheless, marriage is in the air. At the hotel, Stu informs Phil and Doug of his intention to propose to Melissa, and shows them the ring he intends to give her – it's his grandmother's – which made it all the way through the Holocaust with her. Phil is dismayed: 'It's a big fucking mistake. . .she beats him'. 'That was twice', Stu replies 'and I was out of line'. 'Wow', Phil responds 'He's in denial'. It's clear that Melissa is Stu's super-ego: constantly and vigilantly policing his behaviour, his desires and his emotions. Of the four, Stu seems the most strait-laced. So, it is almost a relief to discover that it is Stu who gets married at The Best Little Chapel, yet this is tempered by the 'shock' (and 'joy') that his wife,

Jade, is a sex worker (as a stripper and by providing escort services). Indeed, of the four, it is Stu that is wildest, seemingly freed of his super-ego, his denials and his repressions.

Stu's impulsive, garish and unplanned marriage to Jade is a direct contrast to the carefully choreographed, stylish and thoroughly planned marriage between Doug and Tracy. Yet, of the two, it is Stu's marriage that seems to be the most fun, the most about love. Although the 'style' of the wedding conforms to the image of the tacky Vegas 'quickie', with tawdry suits, a bride in a cowboy outfit, and a bunch of kitsch souvenirs, Stu looks like he is having the time of his life: Stu seems to be in love and, as Phil observes, happy. Jade seems to have released something suppressed in Stu that Melissa simply oppresses. While Stu's marriage is the apparent product of unbridled desire, Doug's seems to be more of a corporate merger between rich families. Yet, it is to the idea of the 'perfect' wedding that Stu returns. The sober Stu is horrified to find that he has married an 'escort': Jade must be found, not just as a step towards finding Doug, but also as a step back from marriage – Stu wants to divorce her as quickly as he can.

In the end, after they have found Doug, Jade tells Stu that what they did was stupid – and a relieved Stu agrees, as Jade hands back Stu's holocaust ring. Sweetly, Stu suggests that they should meet up for a dinner the next weekend, but we know he still has to confront Melissa. During a tirade of accusations from Melissa, Stu finally snaps and (to everyone's relief) publicly dumps her. Here is another contrast between Stu and Doug. As Doug arrives at the wedding, Tracy angrily asks why he is so late and why his skin is so red. Doug says it's a long story, apologises and, importantly, promises never to put Tracy through anything like this again.

Despite appearances, there is a single thought underlying Stu's relationship with Melissa, Stu's marriage to Jade, Phil's relationship with his unnamed wife (even while he greets her passionately when reunited at Doug's wedding), and Doug and Tracy's wedding. This single thought is that marriage is castration and, worse, the end of life. The film displays a consistent fear of women throughout, either that they are directly castrating (e.g. Melissa) or desire for them will overwhelm men and cause them to castrate themselves (e.g. Jade). Indeed, Alan bets Stu that he is not a good enough dentist to pull out his own tooth. Stu wins the bet, by pulling his tooth out in front of Jade. Of course, Stu's display of 'balls' to Jade is also a symbolic castration: a symbol that Melissa instantly recognises as only she is allowed to castrate him. Tracy even has both effects on Doug: his 'wild side' is suppressed by her at the same time as he suppresses it himself.

Of course, male fears of emasculation and castration do not necessarily have to be repressed – and this is why it is important to recognise, along with Freud, that the unconscious is always more than a product of repression and the mechanisms through which repression is maintained. In fact, a whole series of morally repugnant and socially unacceptable thoughts run through the film *The Hangover* – and, of course, this is exactly why it is so funny. What we witness is exactly how gross and stupid men are, and this liberates the audience to laughter. Yet, what the movie does is cover over the disturbing idea that marriage-is-death and women-are-castrators by the reassuring idea of the proper marriage: Tracy and Doug get their perfect wedding; we see that marriage-weary Phil loves his perfectly beautiful wife and cute son; and there is real hope that Stu might now get the girlfriend/wife he really deserves. So, is all well that ends well?

Barely noticeable, the sign on the Best Little Chapel reads 'The Weddings Last'. Stu's marriage to Jade lasts less than a day. As we know, from apocryphal marriages, such as that of Britney Spears to Jason Alexander, Vegas marriages do not last. They are moments of madness that are not meant to leave Vegas. They are not really meant to happen in the first place. It is this tension between what happens and what is not meant to happen that provides the latitude and longitude for the Vegas map of unconscious materials. The key locations are Caesars Palace (suite, poolside and rooftop), the Bellagio and Riviera casinos, the strip club, the car, the vacant lot and the marriage ceremony. Each location does more than simply allow unconscious material to be re-presented, however. Each location is determined by unconscious desires and fears. Thus, hotels and casinos produce an intoxicating mix of pleasure and danger, while strip clubs and weddings hold together a contradictory blend of sex and money, just as cars and vacant lots are sites of possibility and violence. Strangely, when put together, these conform to Freud's description of the key features of unconscious processes.

Conclusion: 'That Shit Will Come Back with You' — Thinking beyond the Spaces of Repression

For Freud, the unconscious processes are distinguished by the privileging of psychic reality; the mobility of psychical investments; ambivalence and the absence of contradiction; and, also, timelessness. It is a mistake to think, however, that this unconscious psychic reality is disconnected from the world. Instead, in this chapter, we learned that psychic reality is entangled with the world in three significant ways: first, the repression of unbearable ideas, affects and experiences is thoroughly worldly; second, repression leaks into the world; and, third, repressed content becomes distributed in the world and recursively acts back on the individual.

In the discussion of Dora, I emphasised the imposition of patriarchal relationships upon her in ways that forced her to both repress her traumatic experiences while at the same finding ways to give these traumatic ideas expression, albeit in disguised ways. This enables us to see repression as a process that navigates the dangerous waters of internal motivations and external imperatives. More than this, repression also manages difficult, unbearable affects. Sometimes, by keeping them away from consciousness, other times by finding alternative forms of expression. We can think of this as drawing different boundaries between knowing and not knowing, between expression and recognition. This directly connects repression to Rancière's arguments about the distribution of the sensible.

For Rancière, the distribution of the sensible draws a line, for example, between those who can speak and those who cannot. This line cuts through Dora, in different ways, at different points. Let's think about her cough. She cannot speak, as a cough prevents her from speaking: the words literally get caught in her throat. Yet, she can also speak, but Freud is unable to hear what she is saying – a problem we follow into the next chapter. Dora, then, embodies no less than two distributions of the sensible, as an effect of the engagement between patriarchal relationships (through her father, Herr K. and Freud) and her inner thoughts and desires (and her relationship with Frau K.). The distribution of the sensible, then, is less an aesthetic regime than a bodily one; not a single bodily regime, but more than one. The next step in the argument about the radical openness of the repressive function of the unconscious is to follow it into the world. To do this, put crudely, we tracked affective forms that are characteristic of patriarchal relationships (such as the fear that women are castrating) in *The Hangover*.

In *The Hangover*, we saw how the distributive function of the unconscious utilises the external world to find forms for its passions, where those passionate forms disguise or hide the unconscious passions that prompt them. These passionate forms are distributed spatially though the world, such that any object might afford the opportunity for unconscious processes to latch onto them. *The Hangover*, in a variety of ways, displays both the repressive and distributive functions of the unconscious. More than just a set of locations, Vegas is a map of psychical reality: a psychogeography of desires and fears and of knowns and not knowns. Put another way, we should be clear that repressed unconscious material is also geographical: producing and produced by its (only seemingly) external worlds. This significance of the extimacy of space, as Kingsbury terms it, is that it can be astatic, holding or prompting contradictory and ambivalent ideas. As the weddings show, contradictions and ambivalences are not just held, they are flaunted and indeed provide opportunity for the very funniest moments. And, all the while, *The Hangover* represents, at various points, the ways in which the men's desires and fears intensify, or de-intensify as they search for Doug. In particular, we see Stu's wild desires turn into something sweeter, yet running through the same object: Jade (named, after all, after a precious stone). There is one more thing, though.

If the evidence never made it out of Vegas, all would be forgotten. All is not forgotten, *The Hangover* warns. What happens in Vegas gets out. It is not an unrepressed city. As the missing tooth shows, there is evidence (even if it is in the form of something that is not there). So, if Vegas were a map of the unconscious, we would be forced to conclude that the unconscious is the product of the relationship between all kinds of unspoken and unthought ideas, but always filtered through social forms of expression. If these ideas appear to have no spatial form, we should remember that the entire film is devoted to putting lost Doug back into in his proper place: into the ideal marriage – the one that masks the thought that marriage-is-death, that women castrate, that men castrate themselves. That is just common sense. Even while those senses are produced in common by Oedipal structures, they remain as operative now as for Dora.

While *The Hangover* mocks the idea of the wolfpack as a description for a group of men, this term took on an all too real meaning on 7 July 2016. During the *San Fermin* celebrations in Pamplona, Spain, a group of five men raped an 18-year-old woman. The men shared video recordings of the rape on their WhatsApp group, titled *La Manada* – The Wolfpack. On 26 April 2018, the five men were sentenced to nine years in jail for sexual abuse, but not sexual aggression (which would have meant 10–15 years in jail). The case has provoked ongoing protests over what constitutes rape in Spanish law and echoes through Spain's #*YoTambién* movement. This brings us back the alignment of repressive functions and the policing of bodies and the brutality of that policing. For me, the language of the wolf pack in *The Hangover* and Spain is not a coincidence, but rather the expression of a bodily regime that creates a particular structure of knowing and not knowing exactly what men are doing with their desires and fears. The wolf pack names a distribution of the sensible. It names an Oedipal structure of the repressive unconscious. It enables the enactment of that repressive unconscious in the world. What is at stake, in *The Hangover* and in Dora's case, is the aggressive assertion of toxic masculinity, driven by forces of repression, sexual aggressivity and male anxieties about sexuality.

What *The Hangover* presents as funny is deadly serious. Dora survived by refusing the assertion of patriarchal assumptions over her experiences. This resistance to Freud's therapeutic assumptions and assertions did more than force a reconceptualisation of the unconscious. Her refusal, perhaps paradoxically, forced Freud to come to terms with not only the full force of unconscious motivations, but also with the dynamics of unconscious communication. This communicative unconscious is critical for understanding psychoanalysis. For, it is the communicative unconscious that transmits and receives affects amongst people (and other objects), albeit in filtered, contingent and distributed ways.

Chapter Six
The Transference of Affect: The Communicative Function of the Unconscious

Introduction

The last chapter focused upon the repressive and distributive functions of the unconscious. Significantly, the unconscious is entangled and engaged with the external world. This has consequences for how we think about the distribution of the sensible and the assignment of bodies to proper places. We can see that the repressive functions of the unconscious happily align with the social assignment of bodies to proper places. Thus, experiences of the uncanny or of disgust or of fear can be strongly associated with bodies out of their proper place. However, repression can also resist 'the proper', fighting back against the unconscious assimilation of social conventions and taboos. Much the same can be said of the distributive unconscious. As we saw in Chapter 5, Oedipal structurings of the distribution of senses and bodies both act upon, and are acted back on, by the individual, in context, in their spaces and places. This helps us understand how unconscious processes might work in support of the production and imposition of bodily regimes. However, this is not all the unconscious is or does. The unconscious, whether as a set of contents, structures or processes, is more unruly than that. Even more so, if we have a model of the unconscious (and therefore, of subjectivity) that is radically open to the world.

Unconscious processes do not just receive messages from the world around; they also express themselves in the world. This is the communicative function of

Bodies, Affects, Politics: The Clash of Bodily Regimes, First Edition. Steve Pile.
© 2021 Royal Geographical Society (with the Institute of British Geographers).
Published 2021 by John Wiley & Sons Ltd.

the unconscious, both receiving and transmitting thoughts and affects. Critically, the distribution of the senses and assignments of bodies to proper places only work if those ideas are communicated, not just consciously, but also unconsciously. Thus, understanding the means of sharing (or not sharing) the contents and structure of the aesthetic unconscious becomes a central, rather than a peripheral idea, in our discussion of bodies, affects and politics. So, in this chapter, we focus upon unconscious communication. That said, it is important to understand that the repressive, distributive and communicative functions of the unconscious are imbricated with one another. They do not operate in isolation from one another; indeed, they are predicated upon each other.

To be sure, the communicative unconscious does not communicate overtly or consciously. It does so in disguised and hidden ways. The most familiar form that the communicative unconscious finds to express itself is, famously, through dreams (which I have discussed elsewhere, see e.g., Pile 2005a). However, there are a vast array of other ways that unconscious ideas and affects can be communicated. Indeed, one significant method that we have already witnessed is through bodily symptoms, as in hysteria (see Chapters 4 and 5). But we can also think about daydreams and jokes, creative practices, mundane body language and the like. These forms of communication are not separate from verbal communication or communication through systems of signs more generally. Rather, overt communication and tacit communication are laminated to one another, such that we can find unconscious resonance in all kinds of things.

We saw in Chapter 1 of this book how the fire at Grenfell altered the fabric of the social in ways that enabled previously marginalised, hidden and silenced experiences to be voiced, those voices to be heard and bodies to be seen and (partially) recognised. We saw that, however fleetingly, a community emerged, through an affective politics of anger, frustration, loss and grief. We cannot simply assume that this community was either natural or inevitable. Further, we cannot assume that there was an already existing 'ready to go' politicised community. To be sure, responses to the fire were built out of already existing political – and communal – resources. Other resources were latent; affective resources, not yet mobilised, but nonetheless available.

Thus, we must wonder about the affective preconditions that enable community to emerge and about how affects are communicated. Long-standing work has gone into understanding how politicised communities work through conscious activism (e.g. Pile and Keith 1997; and, Sharp et al. 2000). As we saw in Chapter 1, writers such as Rancière have considered the unconscious forces at play in the building of political community. For him, these unconscious forces are a set of common senses that structure the body and the political. Yet, the mechanisms that enable senses to be held in common are, in general, taken for granted. Even so, there is been a long-standing appeal to the unconscious to explain how bodies and affects are held in common. For example, Gustave Le Bon prefaces his work on *The Crowd* (1895) with a discussion of the unconscious forces that lie behind the formation of crowds (a preface that is not available in all editions of the book). He says:

> Visible social phenomena appear to be the result of an immense, unconscious working, that as a rule is beyond the reach of our analysis [. . .] Crowds, doubtless, are always unconscious, but this very unconsciousness is perhaps one of the secrets of their strength [. . .] The part played by the unconscious in all out acts is immense, and the part played by reason very small. The unconscious acts like a force still unknown. (1895, pp. 30–31)

Gustave Le Bon was not alone in puzzling about the unconscious forces that comprise social behaviours. The problem of unconscious communication has troubled psychology and psychoanalysis since their inception. The most puzzling examples of this form of communication were being generated in a more occult vein, by spirit mediums, clairvoyants and so on. What occult phenomena appeared to reveal was a set of hidden or latent psychic abilities that suggested a whole new world of unconscious forces and acts. In particular, it suggested that it was possible for people to transmit and receive thoughts and images, mentally and remotely. These experiences are highly suggestive of the functions of a communicative unconscious.

This chapter elaborates these communicative functions to show how bodily regimes emerge, not out of stable and fixed contents, but out of indeterminate and overdetermined processes. This, I argue, helps us understand how affects travel, such that they form the resources out of which communities emerge, circulate and crystalise into action, not as a singularity (a common sense), but as feeling of common purpose (built out of indeterminacy and overdetermination). As important, it provides the connective tissue between bodies and affects. Necessarily, then, this chapter is about the mechanism that enables affects to travel between bodies (a problem I outline in the next section) – and this, in turn, enables us to think about the affective production of politics through the aesthetic (see Chapter 7).

A decade earlier than Gustave le Bon, Frederick Myers was also thinking about the unknown unconscious mechanisms through which feelings and passions could be communicated amongst people (see Hamilton 2009). To describe this mechanism, Myers coined the term *telepathy* (Gurney, Myers and Podmore 1886; see Luckhurst 2002): literally, feeling (-pathy) at a distance (tele-). This is intriguing, not as proof of psychic abilities, but rather as suggestive of the many ways that 'feeling at a distance' might be operative. More than this, early attempts to grapple with the idea of telepathy hint at how the communicative functions of the unconscious become operative through the body. So, I start with telepathy as a form through which psychologists and Freud himself first began to register the creativity and efficacy of unconscious communication. This might seem an esoteric choice, but there are three reasons for this.

First, it allows me to trace a straight line between telepathy, thought transference, transference and counter-transference in Freud's thinking (for recent discussions, see Massicotte 2014; Hewitt 2014; and Wooffitt 2017a). This line of thought, for me, provides the foundations for thinking about unconscious communication. The example, here, is Dora (to extend and deepen the discussion started in Chapter 5). The transference of ideas and affect between people unconsciously is both a problem and a solution in Freud's psychoanalytic method. This issue continues to dog psychoanalysis, in fact. In 1973, Stoller wrote a famous, yet unpublished, paper on telepathic dreams, which prompted a discussion group led by Carol Gilligan and Elizabeth Mayer from 1997 onwards (the Stoller paper was published in Mayer 2001). In this discussion, a variety of telepathic experiences were considered as forms of intuition, unconscious communication and thought transference. Indeed, Mikita Brottman has explored a wide range of seemingly occult phenomena that appear in psychoanalytic settings, including telepathic phenomena (2011). Such telepathic experiences in clinical settings are, for me, indicative of the functions of the communicative unconscious.

Second, what these telepathic experiences – now understood through the idea of unconscious communication – open up is the idea of a radically open subject. That is, a subject that is constantly transmitting and receiving ideas and affects consciously and unconsciously, through both people and objects; objects that include artistic forms as well as everyday objects, words and expressions as well as language itself. This means that psychoanalysis, rather than locating the unconscious in the subject, actually has to ask the question, whose unconscious is it? There is a radical sharing of ideas and affects, yet this does not mean that this is easy, smooth or transparently obvious.

Of course, unconscious communication is tricky both to register and to interpret. This is perhaps why the late nineteenth-century psychologists became preoccupied with phenomena such as talking with the dead, telepathy, clairvoyance and astral travel. Indeed, at the time, the most celebrated psychological text published in 1900 was not Freud's *Interpretation of Dreams*, but Flournoy's *From India to the Planet Mars*. In his celebrated analysis, Flournoy sought to uncover the truth of Hélène Smith's astral visits to Mars. Although her visits to Mars are a focus of the study, Smith also spoke in an unknown language, which she claimed to be Martian, and also to have experienced other psychic phenomena, including telepathy.

To be sure, Flournoy does not discover much in the way of either travelling to Mars or other psychic abilities, such as previous life regression. Analysis of Smith's Martian language finds that it used a substitution code, replacing French words with 'Martian' words. Flournoy, then, was broadly sceptical of Smith's astonishing abilities, except in two significant cases, both of which are telepathic. Against all his other suspicions of Smith's abilities, Flournoy held open the possibility that she had reliably reported telepathic experiences. For me, this is significant because it indicates the moment where we can register, and start to understand, the communicative function of the unconscious – and how it will appear in Freud's and psychoanalytic understanding through notions of trans-ference and counter-transference. So, it is with Flourney (in the section after next) that we will start our exploration of unconscious communication. What this discussion shows, from my perspective, is the sheer variety of ways that Hélène Smith sought to communicate, yet in ways that she herself was either unaware of or only partially so.

Third, *telepathy* as a term is coined to describe phenomena where people seemed to be feeling at a distance. The idea of 'feeling at a distance' contributes to our discussion of the aesthetic unconscious (in Chapter 1) by suggesting that the unconscious structurings of the senses – the production of distributions of the sensible, as Rancière would have it – are formed out of the conscious and unconscious communication of affects that become operative through bodies. This distribution of the senses at a distance is critical. It suggests that affects can be transmitted and received amongst bodies of all kinds at all kinds of distances. This raises a set of questions about the geography of the senses and bodily communication that drive this chapter. Indeed, this discussion leads to the very question of how we understand the senses and what constitutes senses in common (as Rancière might want): thus, clairvoyance disrupts the field of vision, for example, both spatially and temporally; telekinesis hints at the ability to touch and hold at a distance; while telepathy interferes the spaces and limits of the mind-body.

The Problem of Affect Transfer over Distance

One of the critical issues in understanding affect is how it is shared amongst bodies: spatially, socially, physically, psychically. Metaphors, such as transmission and contagion, which seem to provide an understanding of ways that affects are shared in fact generate a series of fundamental questions. How does affect emerge between bodies? How does affect flow from one body to another? How should we think about the space between bodies? In fact, these questions are fundamental to the whole issue of affect – and they have been raised in different ways across the social sciences (and not just in geography; see, for example, Clough and Halley 2007; Gregg and Seigworth 2010; and Wetherell 2012).

Thus, in her book on the transmission of affect (across distance), Teresa Brennan casts the problem this way: 'Is there anyone who has not, at least once, walked into a room and "felt the atmosphere"?' (2004, p. 1). Brennan argues that too little work has been conducted into how it is that people feel the affects of others, as expressed in notions such as atmosphere (see Anderson 2009). For her, the transmission of affect is simultaneously social, psychological and bodily. Affects are social or psychological in origin, but they have body-altering effects in the biochemistry and neurology of the individual (or group). The transmission of affects occurs in two ways: first, through projection; and second, through chemical stimuli.

The term projection is taken from psychoanalysis, and may be broadly seen as analogous to cinema projection. Thus, if Teresa walks into a room with a sense of apprehension, then the atmosphere she experiences will likely be one associated with her apprehension. Similarly, others will also be projecting their affects onto the room (or more precisely people and things in the space of the room). Not all of these affects will, however, become part of the atmosphere of the room: some will be ignored, but some will be recognised, and identified with, by the assembled people. It is out of a kind of tacit sympathising that the atmosphere of a room is built. For Brennan, though, social and psychological origins are not enough to describe or explain the transmission of affects: into the mix, Brennan adds chemical communication, in particular through pheromones and hormones.

The atmosphere of a room is a good place for Brennan to begin her discussion about the transmission of affects, but it is a noticeably circumscribed space – and a quickly problematic one: you have to presume quite a lot to get to the point where you can say that a room has *an* atmosphere, both singular and knowable. This becomes even trickier when you consider other spaces that are commonly held to have the same affect, such as places or cities (see Pile 2005a). It is unclear, also, how groups might respond to, or develop, affectual cues based upon pheromones or hormones over (non-intimate) distance. Brennan is aware of this problem and turns to the issue of 'suggestion' when thinking about the transmission of affect in groups (2004, ch. 3). From this discussion, she learns that the presumption of the suggestibility of groups (as described by Gustav Le Bon and others) underplays their bodily vitality, especially as expressed in aggression and hatred. For her, groups transmit affect through what she calls 'nervous entrainment' (pp. 68–73). Affects are entrained into a group 'feeling' through a mixture of suggestibility, contagion and chemical contact.

The implications of Brennan's arguments have not been lost on geographers. In particular, McCormack has written about the importance of thinking about affects through the body, including thinking through (literal) atmosphere (see McCormack 2006, 2008). Meanwhile, Nigel Thrift has discussed the significance of psychological suggestibility in the manipulation of affects as well as affects as contagious (2004; 2008, ch. 10). Commonly, affect is seen as an intensity that emerges between bodies (following Massumi 2002, ch. 1). Affect is not only registered within bodies, it also produces in bodies different capacities to affect and be affected. From this perspective, affect does not travel between bodies, it is a field that envelopes bodies. Importantly, this begs a simple geographical question. As Phil Crang puts it, in his response to Kathleen Stewart's book *Ordinary Affects* (2007): 'How do ordinary affects vary from place to place, or globally, and how are such variations to be mapped'? (2010, p. 923).

In fact, Crang's question is predicated in an assumption that already we know how ideas and affects emerge between or envelope people – and people and objects.

THE TRANSFERENCE OF AFFECT   119

My suggestion is that the idea of unconscious communication can help explain how affects become 'fields' that envelope bodies, by showing how 'fields' of affect emerge out of communications of all kinds. However, in my understanding, the idea that affects enfold bodies is better understood through the idea of distributions of the sensible, where bodies and affects are connected by conscious and unconscious communication, within and between emergent and diverse bodily regimes. To grasp some of the ways that ideas and affects emerge and are transmitted between people (through objects), I now turn to the intimate relationship that developed between psychologist Théodore Flournoy and spirit medium Hélène Smith – and the not so intimate relationship that quickly deteriorated between fledgling therapist Sigmund Freud and his unwilling patient Dora. My purpose here is less to explore the nature of hypnotic suggestion (partly because I have discussed this elsewhere, Pile 2010b) than to understand Flournoy's avowed and Freud's publicly denied belief in telepathy (see Massicotte 2014).

The belief that the mind somehow travels beyond the body – and indeed can travel great distances – is scientifically established in early psychology (as we will see), with many experiments revealing telepathic experience; and, also in popular culture, especially through theatrical stage acts (on both points, see Lamont 2013). Even so, what emerges between Freud and Dora, and between Flournoy and Smith, does so mainly in the consulting room and in séances – that is, in the kind of space that Brennan uses to prompt us into thinking about the transmission of affect. From Flournoy to Freud, we trace a line of thought that starts with anomalous experiences and leads to the ways that affects and ideas circulate through the bodies of the therapist and the patient. These are instances of unconscious communication, but they also show that this communicative form is highly dynamic, indeterminate and fluid.

The indeterminacy and fluidity of unconscious communication are difficulties that Sigmund Freud became increasingly aware of. In fact, somewhat unpromisingly, he is stunned into silence by them. Even so, in his discussions of telepathy (1921, 1922), and occult experiences (1925, 1933), we can glimpse a covert model of the transfer of affects over distance – one that is unconscious yet non-repressed; one that allows the unconscious mind to reach beyond the body; one that enfolds bodies and spaces. In the final section of this chapter, I explore this model, as it helps us re-situate and extend the presumptions of transmission, suggestion and contagion that currently pervade understandings of how affects flow between bodies by thinking through the unconscious dimensions of communication. I do so not to assert the ontological (or other) superiority of psychoanalytic understandings of the transfer of affect between bodies over distance, but rather to be able to ask further questions about how this happens *unconsciously*. Before we get to Freud, however, we must first understand what is at stake: for this, we need Théodore Flournoy's encounter with Hélène Smith and the truth behind telepathic experiences.

Hélène Smith and Théodore Flournoy on Mars: The Science of Telepathy

Telepathy – as 'distant feeling' – licensed Victorian scientists to investigate a wide variety of nonconscious states of mind and awareness, but especially spirit possession, clairvoyance, hypnosis and trance (see also Holloway 2006). As you may have guessed, the psychological subject responsible for generating the notion of telepathy was 'the medium'. Indeed, most of the leading psychologists seemed to have had their own personal medium to research: William James studied Leonora Piper; Pierre Janet examined Léonie; Frederic Myers drew the Reverend William Stainton Moses and Daniel Dunglas Home into his circle; Sir William Crookes investigated Florence Cook; Carl Jung explored the mediumistic experiences of his cousin Helly; while Charles Richet, along with Pierre and Marie Curie, researched Eusapia Palladino. Many other celebrated mediums were also scrutinized, including Madame Blavatsky, founder of the Theosophical Society, Kate and Margaret Fox, and Mrs Guppy (see, e.g., Conan Doyle 1926; Pearsall 1972; Lamont 2005).

Let us turn to the strange case of Hélène Smith and her scientist, Théodore Flournoy, Professor of Psychology at the University of Geneva. His study was widely celebrated at the time of its publication both for its sceptical investigation of seemingly paranormal powers and also for revealing aspects of the unconscious mind, in particular 'cryptomnesia' (Ellenberger 1970, pp. 315–317). My interest is two-fold: on the one hand, Flournoy's use of hypnosis and the unconscious rapport that developed between him and Hélène Smith (see also Pile 2010b); and, second, his continuing belief in telepathy.

In December 1894, Théodore Flournoy, at the invitation of Professor Auguste Lemaître (of the College of Geneva), attended a séance given by Hélène Smith (real name Catherine-Élise Müller: born 9 December 1861 in Martigny; died 10 June 1929 in Geneva). By reputation, she had 'extraordinary gifts and apparently supernormal faculties' (1899, p. 9). Flournoy was deeply impressed by her. On one occasion, in a series of raps on a table, a message seemingly from beyond the grave conjured up a vision very personal to Flournoy:

> I was greatly surprised to recognize in the scenes which passed before my eyes events which had transpired in my own family prior to my birth. Whence could the medium, whom I had never met before, have derived the knowledge of events belonging to a remote past, of a private nature, and utterly known to any living person? (p. 9)

Flournoy's first thought is that Hélène's abilities remind him of the 'astounding powers' of the internationally celebrated medium Leonora Piper – 'whose wonderful intuition reads the latent memories of her visitors like an open book' (p. 9). Now, he had met someone for himself who had apparently supernormal powers. This was his opportunity to study these powers, to determine whether spirit manifestations, clairvoyance, telepathy and the like had any genuine substance or not. Over the next five years, Théodore Flournoy attended many séances, as Hélène Smith freely consented to participate in his study. In this period, Hélène's mediumistic practice changed. She graduated from table-rapping and visual and auditory hallucinations to the higher plane of spirit possession and total somnambulism (a hypnotic state). So deep were these trance states that, in a series of experiments in 1895, Flournoy and others were unable to wake Hélène. For Flournoy, her states of trance were:

> thoroughly identical with those that may be observed in cases of hysteria (which are more permanent), and those that may be momentarily produced in hypnotic subjects by suggestion. (1899, p. 12)

Already we can see that Flournoy is beginning to associate Hélène Smith's mediumistic trances with hysteria and hypnosis. Here, Flournoy is drawing on orthodox understandings that saw similarities in the mental processes that create experiences of hypnotic states, suggestion and hysteria (see Didi-Huberman 1982). More than this, Flournoy was aware that his presence in the séances was exerting an influence on her mediumistic performances.

For Flournoy, Smith's performances were not simply fictions that she would conjure up under specific circumstances, for he believed that Hélène was also suffering from what is popularly known today as Dissociative Identity Disorder or Multiple Personality Disorder (see also Platt 2010). For Flournoy, the roots of Smith's visits to Mars, for example, probably extended far back into Hélène's childhood. He argued that she had forgotten these experiences in such a way that, when she dredged them up from her subliminal mind, she could no longer recognise them as from her childhood. Flournoy called this 'cryptomnesia': memories that are so hidden that they are no longer recognisable as such. In Flournoy's understanding (much like Freud's; see Chapter 4), although Smith had forgotten her childhood experiences, her subliminal mind continued to work on and with them – producing, over time, ever more elaborate iterations of each performance. In her trances, then, Smith did not simply dig up the fossil relics of her childhood, her subliminal mind 'took over' her body and her consciousness (Flournoy 1899, p. 89).

There is a complex model of the mind, here: it is split horizontally into the subliminal, the subconscious and the supraliminal (drawing heavily upon Myers' model of the mind; see Myers 1903); it is also split vertically, into different personalities. Flournoy became aware, also, that he was unwittingly influencing Smith's performances. In one séance, for example, Smith has a vision of a woman with black hair (1899, p. 176). Smith looks in ecstasy at the woman, as she herself moves towards Flournoy. Though Smith seems neither to see nor hear Flournoy, she gestures towards him with her hands. Flournoy changes his seat, but Smith's attention, seemingly following the movements of the black-haired woman, always turns to him. Indeed, what Flournoy's study ultimately seemed to prove (to him) was the extent to which there was an unconscious transfer of affect between himself and Hélène (Ellenberger 1970, p. 316).

It is clear from the outset that Flournoy himself is struck by Hélène Smith; she, on the other hand, produced ever greater marvels for Flournoy; together, these can easily be read as evidence of mutual attraction and seduction. Indeed, for Ellenberger, the rapport between Flournoy and Smith is clearly psychosexual in nature (1970, p. 316). So, the relationship was telepathic to the extent that the rapport between them involved the involuntary telegraphing of wishes and desires between them. The pair of them danced around each other, as it were, producing for each other a subliminal romance: The Professor and The Medium. Yet, for Flournoy, the rapport between them was not telepathic, but a feature of the practice of hypnosis – which produced hidden erotic dynamics and a high degree of suggestibility. Instead, Flournoy found evidence of telepathy elsewhere in his experiences with Hélène Smith, albeit only a little (p. 267).

By going through Flournoy's study of Hélène Smith's mediumistic performances, I have so far tried to illustrate two things: *first*, that there was an unconscious transfer of affect between Smith and Flournoy; and *second*, that this unconscious communication of affects is two-way and operating through in a variety of media: performances, fantasies, table rapping, and so on, but primarily through the body, in comportments, gestures, voice, facial expressions and the like. Despite the astounding powers of the mediums, Flournoy remained entirely sceptical about all occult phenomena – all, that is, but two: telepathy and telekinesis. Both, we must note, involve the distribution of the senses at a distance.

Hélène Smith and the Distribution of the Senses: Telepathy and Telekinesis

For Flournoy, Hélène Smith was a psychological subject, not an occult one: her occult experiences were psychological in origin, not supernatural. Intriguingly, Flournoy remained convinced that telepathy and telekinesis were real. On telekinesis – the ability to influence objects at a distance – Flournoy cites Sir William Crookes experiments with Daniel Dunglas Home (Flournoy 1899, p. 231). Crookes had shown that Home was able to move, and to alter the weight of, (heavy) bodies by mental effort alone (see Crookes 1874, pp. 88–89).

Significantly, for Flournoy, telepathy was not simply real – it was also necessary:

> One may almost say that if telepathy did not exist one would have to invent it. I mean by this that a direct action between living beings, independent of the organs of the senses, is a matter of conformity to all that we know of nature that it would be

hard not to suppose it *a priori*, even if we had no perceptible indication of it. How is it possible to believe that the foci of chemical phenomena, as complex as nervous centres, can be in activity without giving forth diverse undulations, x, y, and z rays, traversing the cranium as the sun traverses a pane of glass, and acting at a distance on their homologues in other craniums? It is a simple matter of intensity. (Flournoy 1899, p. 235)

Drawing on the notion of rays, Flournoy wonders about the materiality of telepathic relations. The analogy of the sun shining through glass gives him confidence that telepathy could be grounded in physics: that is, in physical reality.

Flournoy was, consequently, convinced telepathy was a psychic reality. Indeed, he had detected some telepathic episodes in his relationship with Hélène Smith – although Flournoy believed she had closer telepathic relationships with others, especially Monsieur Balmès. Thus, for example, while gazing into a crystal ball, Smith saw Monsieur Balmès (who had been called suddenly to Geneva) with a friend, above their heads was a pistol. On his return, Balmès reported that a friend had indeed offered him a pistol. Crucially, Flournoy does not attribute this to Hélène's ability to actually see Balmès at a distance. Instead, he describes this as 'anticipated telepathy' (Flournoy 1899, pp. 240–241). Flournoy reasons that Balmès probably anticipated that his friend would offer him a pistol. This anticipation was telepathically communicated to the unwitting Hélène Smith. The anticipation then manifests itself in Smith's vision.

Flournoy believed that, in some psychic states, Hélène Smith was able to receive and transmit impressions telepathically – and that this ability enabled actual clairvoyance (understood as telepathy that can touch events in the future). From this perspective, there are three important features of telepathy. *First*, proof of telepathy is not to be found in Zener card ESP experiments. It is rather experienced in *nonconscious* states of mind, especially in dreams and trances and while under hypnosis. *Second*, telepathy is 'supernormal' not 'supernatural'. While some may be more attuned, under specific circumstances, to telepathic subliminal communication, it is a *normal* human ability to both receive and transmit subliminal impressions or suggestions telepathically. *Third*, there appeared to be some ability for the mind to travel across space and time: that is, to view events not only at a geographical distance, but also in the past and future.

Flournoy's findings matched almost exactly those of his contemporaries. Pierre Janet had uncovered the relationship between telepathy, suggestion and hypnosis in his experiments with Leonie (Myers 1903, pp. 108–110). Myers' experiments with Malcolm Guthrie had revealed a remarkable ability to communicate images telepathically to others (Myers 1903, pp. 133–140). Myers suggested that there was a 'community of sensation', underpinned by both a telepathic *rapport* between Guthrie and others, and also, perhaps, by the harmonisation of brain waves – that is, by a literal meeting of minds over distance. Meanwhile, Mrs Sidgwick had reported, what she called, 'travelling clairvoyance' during a séance:

in a trance state, Jane, the wife of a Durham miner, was able to 'travel' to other places. Thus, for example, on one travel, Jane described a red house, with a laburnum tree. The door knocker, she said, was made of iron not brass. Inside the house, Jane saw a man at a table, reading papers, drinking brandy. The following morning, the accuracy of her description was confirmed in every detail (Myers 1903, pp. 117–118).

What these instances of telepathy and telekinesis demonstrate is a range of experiences and phenomena through which unconscious communication might be registered. Rather than think of these events as real in themselves, we can see how they become affectually meaningful through their interpretation (along these lines, see Pile, Bartolini and MacKian 2019). This is not extraordinary or surprising. Everyday life is littered with anomalous and difficult to explain experiences. Such events are normal. Around the end of the nineteenth century and into the early twentieth century, telepathy was not simply a fantasy; it was an established phenomenon – a *necessary* concept to explain 'community of sensation': the experience of shared affects between people at a distance. This community of sensation relies upon the sharing of ideas and affects between bodies, both human and nonhuman (see also Wooffitt 2017b). Indeed, it is easy to see parallels between the ideas of community of sensation, fields of affect and the distribution of the sensible.

The promise of the mediums was that they would be able to open up a latent ability in humans and to answer the biggest question 'what happens when we die'?, whether through talking with the dead or with each other mentally or through moving bodies in mental and material space or through manipulating physical objects in space. Yet, these anomalous phenomena were being interpreted in a much more ordinary and mundane way by Freud, as about the transfer of affects and ideas, unconsciously, through the body.

Feeling at a Distance: What Would Freud Know?

In his paper on dreams and telepathy (1922), Freud finally observes:

> Have I given you the impression that I am secretly inclined to support the reality of telepathy in the occult sense? If so, I should very much regret that it is so difficult to avoid giving such an impression. In reality, however, I was anxious to be strictly impartial. I have every reason to be so, for I have no opinion: I know nothing about it. (p. 86)

Freud is careful to choose his words. He wishes to be impartial, has no opinion, yet knows nothing about it. 'It', to be clear, is telepathy in the occult sense. But has Freud left a gap here: what of telepathy in the non-occult sense? Elsewhere, Jan Campbell and I have argued that psychoanalysis is covertly reliant on a non-occult understanding of telepathy (Campbell and Pile 2010). Here, however, my focus is on how Freud might provide a model for affect transfer at a distance: *unconscious communication*. In this model, not only do affects *circulate* between people and their objects, but affects and thoughts are also *intimately entangled* (see Campbell 2007).

Perhaps the most curious thing about Freud's two papers on telepathy (1921, 1922) is the lack of any examples that might be obviously telepathic. Thus, two cases in 'Psychoanalysis and Telepathy' (1921) concern prophecies that did not come true. The third case involves observations about graphologist Rafael Scherman, who was able to make inferences about a person's character from their handwriting. In 'Dreams and Telepathy' (1922), there are several cases of dreams that anticipated future events. There is also a case where a daughter seems to have communicated with her father – telepathically, while he slept, in a dream – the unexpected news of the birth of twin grandchildren (pp. 74–76). Indeed, these experiences continue to be reported. For a recent discussion of the enigma of telepathy in dreams, see Eshel (2006) and also Mayer (2001). Significantly, Eshel interprets telepathic experiences as an expression of the unconscious communication that occurs between therapist and patient. To find out what is at stake in these experiences, let me take Freud's clearest example of telepathy, which involves the relationship between Freud, a patient of Freud's, the patient's (ex) lover and the graphologist Rafael Scherman (1921).

Rafael Scherman is rumoured to have invented the term *graphology* in about 1915. His approach, like that of Curt Honroth, was to read handwriting for what it revealed about the emotional state of the writer. This reading would then allow Scherman to deduce the character of the writer. One technique he used was to look for 'lapsus calami': that is, slips of the pen. A technique, it ought to be observed, similar to Freud's looking for 'lapsus linguae', or slips of the tongue. Freud claimed no direct experience of Scherman's 'amazing feats' (1921, p. 66), but a patient had. After ending a love affair, the patient, a young man, had attempted suicide and sought treatment from Freud. At the same time, the young man was also consulting Scherman, giving him samples of his lover's handwriting. From the lover's letters, Scherman concluded that she was at the end of her emotional tether and that she was 'certain to kill herself' (p. 67). Later, in fact, she did attempt suicide.

Freud sees Scherman's successful prediction as a consequence of neither his graphology nor his clairvoyance. Instead, Freud argues, the prediction arose from his patient's 'secret wish'. Scherman, somehow, had picked up the young man's secret desire for his lover to die. Freud, however, does not say how it was possible for the patient's unconscious wish to be communicated to Scherman without either of them knowing. Put another way, what Freud avoids saying is that the patient's unconscious wish is communicated *unconsciously* to Scherman – *and also* to his lover (and, indeed, to Freud as well). Of course, the young man's 'secret' may not have been conveyed either unconsciously or telepathically in either situation. However, Freud's insistence that the wish is secret – that it is *repressed* – indicates that he does indeed believe his patient unconsciously telegraphs his wish to both Scherman and his lover. That is, Freud believes implicitly that unconscious wishes, as both an affect (the death of love) and an idea (the death of a lover), can be conveyed between people, at a distance, without anyone being aware that this communication is happening. Rather than calling this 'telepathy' (thereby invoking its occult connotations), however, Freud prefers the term 'thought transference'. He insists:

> I have nothing to say about the rest of the wonders of occultism. I have already publically admitted that from the occult viewpoint my own life was a singularly barren one. Perhaps the problem of thought transference will seem rather trifling to you in comparison with the great world of occult miracles. Yet consider that even this hypothesis already represents *a great and momentous step* beyond our present viewpoint. (1921, p. 68, emphasis added)

A great and momentous step? The step forward that Freud makes is to be able to jettison the occult aspects of the idea of telepathy, while being able to hold onto the non-occult aspects by using, instead, the notion of 'thought transference', and later 'transference'. As Jan Campbell has pointed out (2006, 2007), for Freud, thought is neither conscious nor affect-free. Thus, as she argues, affects and thoughts are in some way connected, albeit opaquely and through chains of association and disassociation, as in dreams. Transference is, of course, a cornerstone of psychoanalytic and psychotherapeutic thought and its unconscious dimensions are now well known (for recent reviews, see Platt 2009; or Redman 2009).

Importantly, for Freud, transference and counter-transference always involved the unconscious transfer of affects, as Campbell has shown (2007). What Freud's notion of 'thought transference' does, however, is describe a situation not only where affects circulate between bodies, unconsciously, but also where these affects can emerge into conscious thoughts, even though these might not be recognised as such. Thus, transference relies upon a covert model of unconscious communication between people and their objects: whether between lovers, letters and graphologists, or between psychoanalysts, their patients and their worlds.

Intriguingly, the notion of the unconscious that underpins this 'telepathic' communication is not just the repressed unconscious with which Freud is most closely associated. Simply put, repression would prevent the unconscious mind from letting repressed material leak into the world. So, for unconscious communication to take place, Freud reminds us that there is also a nonrepressed side to the unconscious mind (see Campbell and Pile 2010). The significance of this is that the repressed unconscious and nonrepressed unconscious are open/closed, bounded/unbounded, distributed in different ways, allowing for highly variable transferential experiences – some of which may be so fantastic as to appear occult.

Publicly, Freud was sure to side with science (transference) against séance (telepathy). Privately, he is clearer about holding to the idea of telepathy. In a letter dated 8 May 1901, Freud writes to Wilhelm Fliess:

> Your letter gave me by no means the least pleasure, except for the part about magic, which I object to as a superfluous plaster to cover your doubt about 'thought reading'. I remain loyal to thought reading and continue to doubt 'magic'. (1887–1904, p. 440)

Their difference of opinion heats up. On 7 August 1901, in an argument about Fliess' wife's relationship to Josef Brauer, Freud's anger shows:

> In this you too have come to the limits of you perspicacity; you take sides against me and tell me that 'the reader of thoughts merely reads his own thoughts into other people', *which renders all my efforts valueless*. (1887–1904, p. 447, emphasis added)

We have come full circle: Teresa Brennan argues that the transmission of affect is based, psychologically, upon the mind projecting its thoughts into other people. For Freud, such a view, whether expressed by Fliess or Brennan, renders his psychoanalytic project valueless. To be sure, this is not because the idea of affective projection undermines psychoanalysis, but because it is insufficient. Projection must be accompanied by introjection. Projection is only one psychical mechanism (of many) that enables an individual to pick up the affect or affects in a room. In this chapter, we are focused on unconscious communication. This is because it introduces a sense of the transmission and reception of affects and ideas amongst the people in the room. More than this, it presumes that this communication can occur through all kinds of media, including the body, but more than the body. It can involve body language, but equally the architecture of the room itself. It can include the language spoken and also the way it is being spoken. It can include people-as-objects and objects and so on.

Psychoanalysis not only presumes unconscious communication, its therapy is entirely based upon bringing its structure and content into consciousness so that it can be worked on and worked through. Indeed, psychoanalysis' emphasis on dreams, fantasies, slips of the tongue, bodily symptoms, repetition are all about witnessing the expression of unconscious material, as mediated by the communicative unconscious. These experiences are 'noises off' the stage of ordinary life, but there are also experiences that seem extraordinary and unaccountable. These experiences include finishing other people's sentences, knowing when you are being looked at, thinking about someone then receiving a call (email, message, etc.) from them, knowing what someone is going to say next, and so on.

Classic Freudian terms such as transference and counter-transference can be seen to have their feet in such anomalous experiences (see Pile 2010b). Yet, paying attention to such marginal and anomalous experiences is Freud's way to capture the psychodynamics of unconscious communication in the clinical setting. In this, it shares much with the psychology of Flournoy. As both Freud and Flournoy were seeking to attune both to the transfer of affect and also to the capacities of the mind and body in different states of consciousness. Here, rather than the psychic or the spirit medium, the hysteric becomes the test case of unconscious communication and the distribution of the senses. For this reason, we return to the case study of Dora, where we discover her strange ability to communicate with Freud – without using words. Indeed, arguably, he was not (even) able to hear her words.

From Telepathy to Thought Transference: Dora and What Freud Does Not Know, but Knows He Needs to Know

In this section, I explore the ways that Dora and Freud unconsciously communicate with one another (for background to Freud's Dora case study; see Chapter 5). In 1900, Dora spent 11 weeks with Freud, undergoing therapy for hysterical symptoms such as coughing and breathing difficulties. For Freud, the key to treating these symptoms lies in understanding Dora's relationships with two men: her Papa and a family friend, Herr K. The course of therapy does not go well for Dora. Indeed, the conscious communication between her and Freud is not promising for Dora, as Freud continually asserts his interpretations over her experiences. It is perhaps this, above all, that intensifies Dora's conscious resistance to therapy – and forces her to abandon Freud. Nonetheless, Dora's fierce resistances to Freud are also unconscious. They go together. For me, this indicates that conscious and unconscious communication are imbricated in one another. More than this, they are similarly structured: they are incomplete, indeterminate and overdetermined, and mutually constitutive. One is not structured like the other, they each structure one another. The imbrication of psychical processes occurs in physical space, through the body, in context. Mental processes are not just modelled on the body (see Chapter 3), they play out through the body. In specific social and spatial settings. Thus, Freud's interpretations would be registered by Dora both consciously and unconsciously. For Dora, the experience of Freud's nascent psychoanalytic therapeutic practice was far from supportive and understanding.

Troublingly, throughout his case history of Dora, Freud seems far from being gentle or deft in his handling of her sexual experiences and traumas. Worse, he admits to playing tricks on Dora to solicit 'hidden' material, as when he places matches on his desk (p. 62). Arguably, his interpretative interventions say more about his fantasies than Dora's realities, as when he asserts that Dora must have felt Herr K.'s erect penis in the shop (p. 22), which enables him to conclude that Dora's coughing and breathing problems must have something to do with oral sex (p. 37), or when he connects Dora's intimacy with Frau K. to a repressed desire for Frau K. (p. 47). There appears to be a rush to judgment on Freud's part, in his eagerness to prove both his theory that hysteria has its origins in sexual trauma and also the therapeutic importance of his dream analysis. Perhaps it is for this reason that Dora abruptly ends her analysis? Freud wonders:

> Could I have kept the girl in treatment if I had found a part for myself to play, if I had exaggerated the importance of her presence for myself, and shown her a keen interest, which, in spite of the attention caused by my position as a doctor, would have resembled a substitute for the kindness she longed for? I don't know. (1905, p. 95)

If Freud had found a part for himself to play? Later in the case history, Freud begins to suspect the exact opposite – that, in fact, he already had a part to play – he had become, for Dora, a man much like other men; a man much like Papa and Herr K. 'But I ignored the first warning', Freud bemoans:

> telling myself that we had plenty of time, since no other signs of transference were apparent, and since the material for the analysis was not yet exhausted. So transference took me by surprise, and because of whatever unknown factor it was that made me remind her of Herr K., she avenged herself on me, as she wanted to avenge herself on Herr K., and left me, just as she believed herself deceived and abandoned by him. In that way she was acting out a significant part of her memories instead of reproducing them in the cure. (1905, p. 106)

The psychoanalysis of Dora involves Freud, basically, conducting two kinds of investigation. On the one hand, he scrutinises Dora's sexual experiences and traumas, from her earliest childhood memories up to her relationships with Papa and her mother, with Herr and Frau K. – with forensic detail and gynaecological objectivity. On the other hand, Freud utilises his theory of dreams to track Dora's conscious and unconscious worlds of meaning and affect. In this mode of psychotherapy, the patient is rendered passive. Their role is simply to provide evidence of the disease, for it is the doctor that must do the work of interpretation and therapeutic intervention. Or so it appears to Freud, in this moment. Dora forces a different approach.

Significantly, this therapeutic practice does not involve a 'talking cure' (as Anna O. termed it: 1905, p. 38), but listening and translation. First, Freud has to hear what the patient is saying: this he calls evenly suspended attention, as it involves listening to what patients are saying beyond what they are consciously expressing. That is, therapists must attune themselves to the patient's unconscious communications. Then, Freud must translate unconscious meanings into conscious ones. As with any translation between any language and any other, this translation is fraught with well-known and obvious difficulties. It is made no easier by the fact that the unconscious is only like a language in a few respects, and its contents are more like a jumble of dynamically interacting memories, ideas, images, affects and so on than a dictionary of definitions (although dictionaries themselves, not by accident, also resemble jumbles of ideas).

Through listening and translation, Freud brings to consciousness the patient's disguised and hidden unconscious thought processes. But this only describes what might be going on in Dora's repressed unconscious: the sea of unwanted contents that she attempts to keep from drowning her. He can do nothing with this unless he can alter or shift these unwanted contents, for this he needs to be able to 'speak' to Dora's unconscious. This requires an ability to provide therapeutic interventions that work consciously and unconsciously. Covertly, Freud builds his understanding of what is required on extant models of telepathy. However, to avoid accusations of occult thinking, Freud introduces the term *transference*. Transference is not the direct communication of unconscious thoughts to others, but rather describes shifts in affects from one person to another. With Dora, we start to see the development of Freud's understanding of the communicative function of the unconscious, albeit initially as a one way process (from patient to doctor). Yet it is Dora that forces Freud to rethink the dynamics of transference, which, he admits, he not only missed or ignored, but he also did not fully appreciate its use therapeutically (if at all).

As I have said, Freud's interpretations in this case study appear to side, very strongly, with the work of culture. A 25-year-old girl who is trapped by on older man, who then forces a kiss upon her, he says, ought to have responded as if this was 'an occasion for sexual excitement' (p. 21); he asserts that women are prone to a more intense jealousy than men (p. 50); he imagines Dora 'felt not only the kiss on her lips but also the pushing of the erect member against her body' (p. 22), without giving any reason for thinking that this ought to have been the case; he imposes interpretations upon Dora, as when he tells her there is 'no doubt' that her coughing was designed 'to turn her father away from Frau K.' (p. 32) and a represents 'a situation of sexual gratification *per os*' (p. 37), that is, that Dora's coughing is associated with oral sex; when Dora says 'no' or 'I *knew* you'd say that' to Freud's interpretation, he responds by taking this as proof that his interpretations are correct (p. 46, p. 60); he shares knowledge of what Frau K. allowed Dora to read with Dora, so that she was clear that the adults were communicating about the most intimate aspects of her life behind her back (p. 19); and Freud even puts words in Dora's mouth when he seeks to translate an unconscious thought into a conscious form, thus: 'Since all men are so appalling, I would prefer not to marry. This is my revenge' (p. 107).

Indeed, the outcome of the analysis seems to have proved to Dora that 'all men were like Papa' (1905, p. 71) – including Freud. Too late, Freud recognises that, somehow, he has become embroiled in Dora's internal worlds. For her, Freud is just another of the 'so appalling' men. Although Dora's abrupt breaking off of analysis comes as a total surprise to Freud at the time. He belatedly comes to realise that it made perfect sense for Dora. She avenged herself upon him, just as she sought to avenge herself on Papa and Herr K. So, in this regard, transference refers to the way that Dora came to see Freud as if he were Papa, Herr K. and, indeed, appallingly 'all men'. This explains why Freud wonders how therapy would have proceeded if he had either resisted Dora's transference, or indeed played into it – by becoming even more like Papa and Herr K. Indeed, transference is fundamental to psychoanalytic therapy:

> If one goes into the theory of analytical technique, one comes to the understanding that the transference is something that it necessarily requires. In practical terms, at least, one becomes convinced that one cannot by any means avoid it. (Freud 1905, p. 104)

Sure, Freud says, it is easily possible to learn the method of dream analysis and to translate the dream's unconscious thoughts and memories into conscious interpretations (p. 104). But transference makes the process of psychoanalytic interpretation fraught with uncertainty: it is not that the patient begins to treat the doctor as if they were a pre-existing character in their psychodrama – as when Dora appears to 'love' Freud *as if* he were Papa or Herr K. (p. 105). The real problem is that the doctor himself is also treating the patient 'as if' they were a character in their own psychodrama. Perhaps this is why Freud imagines erect member? Has he fallen in love with his 'intelligent' patient with 'agreeable facial features' (p. 17)? As Freud admits, transference effectively reduces the analyst's work to guesswork (p. 104). This is, perhaps, psychoanalysis' saving grace: it is necessarily humble and uncertain in the face of not only the patient's symptoms, but also of the process of analysis itself. It is Freud's failure to fully appreciate the dynamics of transference – that is, to fully appreciate what Dora is telling him, consciously and unconsciously, and therefore how he is complicit in re-traumatising her – that causes the analysis to break down. This is what he did not know at the time.

There are three aspects to the puzzle of transference that shadow psychoanalysis. First, it is not clear what unconscious thoughts the patient is transferring onto the doctor. Second, the exact same is true of the doctor, no matter how critically reflexive they are, unconscious thoughts remain illusively unconscious. Third, this is not like a cinema with an opaque screen in the middle and two different films being projected on each side of the screen, enabling one film to be seen behind the other, but where distinguishing between the films become difficult (if not impossible). The entanglement of conscious and unconscious thoughts is critical to understanding unconscious communication in both clinical and everyday settings. Whether unknowingly or not quite knowingly, people's unconscious thoughts can be transmitted and received unconsciously and those thoughts can be worked upon unconsciously (a model best demonstrated in Freud's later discussions of telepathy: 1921 and 1922). These processes are not, Freud insists, unique to the clinical setting, but they are its greatest asset.

Psychoanalytic cure does not create the transference, it only reveals it, as it does other phenomena hidden in mental life [. . .] The transference, destined to be the greatest obstacle to psychoanalysis, becomes its most powerful aid if one succeeds in guessing it correctly on each occasion and translating it to the patient. (1905, p. 105)

If one succeeds in guessing it. And can translate it. A big 'if' and a bigger 'and', as Freud admits:

I did not succeed in mastering the transference in time; the readiness with which Dora put part of the pathogenic material at my disposal meant that I neglected to pay attention to the first signs of the transference, which she prepared with another part of the same material, a part that remained known to me. (1905, p. 106)

It is not that Freud misses the way that Dora compares him to her Papa, or that the ways Dora warns herself about the bad intentions of Herr K. resemble a caution against Freud (pp. 105–106). It is more that he neglects to use these intuitions in the course of the therapy. Dora is troublesome: she refuses Freud many times over – and, ultimately, despite Freud's claim that the therapy had partial success, ultimately the case history does not work as convincing proof of either dream analysis or of psychoanalysis. However, what Dora has given psychoanalysis cannot be underestimated – not because she is somehow a 'special case', but because she demonstrates what is ordinarily going on. Dora has shown Freud a series of unconscious processes that reveal how unconscious thoughts become manifest in dreams and through her body – as if she has given him garlands of flowers. But these flowers are stretched out on barbed wire. Both Dora and Freud have to be careful when grasping at the garlands of unconscious thoughts, for it is never clear whose thoughts they are, nor indeed why they are where they are, and why they take the form they do.

Moreover, a too narrow an understanding of the body and the mind – where the mind is simply a product of the fleshly mechanics of the body; and, where the mind stops at the limits of the material body – is unable to account for transferential experiences. Put another way, transferential experiences implied, at least, an open, distributed and unbounded side to the mind, especially in its nonconscious states. Indeed, transference and counter-transference suggests two-way communication. But, more than this, it implies an enfolding of bodies and spaces. This radically redistributes our understanding of the senses, by connecting the senses to the communicative function of the unconscious: sight, hearing, touching, feeling, smelling, taste and so on are no longer simply the product of physical organs (although they are also that), but worked over many times by psychical processes. Significantly, this makes the mind-body an unreliable witness to (or witnessing point for) the distribution of the senses.

Conclusion: Affects, Bodies and Unconscious Communication

In its origins, *telepathy*, as a term, was part of an expanding lexicon describing the ways affect moves between bodies: starting with the coining of the term sympathy in the 1570s up to the term *sociopath* in the 1930s. In this context, telepathy is best understood as 'feeling at a distance' or 'distant feeling' rather than as mind reading and the like. Understood this way, telepathy is one concept (amongst many) through which the ebb and flow of affect between bodies can be grasped. Indeed, Freud generated his own terms for this ebb and flow of affect: transference and counter-transference. What is significant about this is that the ebb and flow of affect is not one-way (say, from patient to therapist). Indeed, it is not just two-way, like a telephone conversation, but more like a group call, in which the patient, therapist and a whole cast of characters and social settings make their appearance. In Dora's case, there was Papa, Herr K. and Fraud K. and her unnamed mother: not just the consulting room, there was the lake and the reading room. This makes the communicative function of the unconscious profoundly social. It also opens up the subject to their social world in radical ways.

There three key lessons to be drawn from the discussion of unconscious communication in this chapter.

(1) Like the notions of transmission, contagion and suggestion, telepathy involves the carrying of affects from one mind to another. Indeed, in its occult guises, telepathy describes the ways in which the transfer of affects might be controlled and channelled or open and unbounded. In its psychoanalytic conception, telepathy undergoes conceptual transformation. This transformation is in part prompted by Freud's desire to avoid psychoanalysis being tainted by accusations that it is mystical and occult. However, the emergence of the idea of unconscious communication laminates unconscious thoughts and ideas to ordinary and prosaic communication. This renders communication constitutively overdetermined and indeterminate: indeed, rather than exceptional, we might think of the experience of being misunderstood or of being misrecognised as profoundly ordinary. Communication works, not because it is clear and obvious, but because it is constantly seeking to overcome the contradictions and ambivalences contained within it. Thus, picking up the affect of a room is never straight-forward, as Brennan suggests. Rather, it is indeterminate and dynamic. And multilayered, as the distribution of affects works through the communicative unconscious as well as the repressed and distributed unconscious (and also, we can add, a noncognitive unconscious, which drives breathing, heart rate, etc.).

(2) In recent times, efforts to understand subjectivity have tended to rely on either a bounded or external model of the body and mind: bounded, where personhood is located within an integrated and coherent body (as in phenomenology); external, where it is determined by forces that lie beyond the body (as in social constructionism). Unconscious communication instead shows, on the one

hand, that affect cannot be thought of as if it were a (social or material or affectual) soup in which every body swims; and on the other, that affects, feelings and emotions are not simply the product of responses to the stimuli of the external world (see also Pile 1996). Further, this model demonstrates that unconscious communication between bodies is transmitted and received under specific circumstances. It may be normal, and constitutive of everyday interactions and encounters, but it is not either an always 'on' or an always 'off' switch. Particular psychical structures (such as the strength of an individual's repression, or their masking of unconscious dynamics by rationality) and particular psychical states (from dreams to trances to daydreaming to consciousness) make people, bodies, more and less attuned to marginal and anomalous experiences, including telepathic ones.

(3) Unconscious communication presents a model for understanding the ways that feelings are transmitted or picked up between people and amongst people and the external world. Further, this model also implies that the capacities of minds and bodies have, at least, different spatialities – some of which coincide, others of which do not (see also Pile 2009). I have argued that psychoanalysis is well placed to describe unconscious communication between people, between people and their objects, precisely because the unconscious is seen as dynamic, multiple *and* distributed. Psychoanalytic concepts such as thought transference, transference and counter-transference suggest that the emergence of affects and thoughts in any particular setting cannot be assumed, but must be accounted for. This account must take into account not only people's consciously expressed understandings of, and feelings about, that situation, but also the presence of unconscious psychodynamics and unconscious communication.

This is exactly the kind of interpretation that we saw Fanon give of his encounter with the small boy (in Chapter 2). The encounter was not only about what the boy said, but about the unconscious hinterland of affects and ideas that came with his exclamation. Fanon instantly picked this up, unconsciously. Accounting for this requires not simply an attendance to these dynamics of their encounter, however, to this must be added the variable topographies of the encounter. Unconscious communication, then, asks us to understand how specific affects and thoughts emergence and circulation in specific situations, in their specific geographies. This does not, however, make the social disappear. Rather, it infuses the social with the transmission and circulation of affects and ideas.

Or, put another way. Throughout this book, I have sought to show that bodily regimes do not 'add up'. Rather, bodily regimes are comprised out of various schema, both consciously and unconsciously, that are intensely and ruthlessly policed precisely because they do not quite fit together into one seamless, integrated, coherent whole. Indeed, objects can become sites that crystallise affects and thoughts that do not quite 'add up'. Moreover, objects can be produced that deliberately create dissonance between the various affects and social relationships that seek to comprise them. Politics, then, seeps into the fissures between schemas and regimes – as much as out of those rare times when there is an 'outside' to a regime. Bodily regimes, then, can be deliberately called into question by objects – and aesthetics, in Rancière's terms – that forces these fissures into view, both conceptually and affectively. So, in the next chapter, we explore how this is achieved in the art work of Sharon Kivland, who makes strangely, distrubingly beautiful objects – which call into question the reliability of the aesthetic unconscious in the production of a singular, coherent aesthetic regime.

Chapter Seven
Crazy about Their Bodies: The Art-work of Sharon Kivland and the Politics of the Female Body

Introduction: The Aesthetic and the Unconscious

On 5 October 2014, Sharon Kivland's exhibition, *Folles de Leur Corps* (Crazy about their Bodies), opened at the Café Gallery Projects in southeast London. The gallery is known for hosting exhibitions of contemporary art. Designed by the Bermondsey Artists' Group, it offers a purpose-built space for artists. Sharon Kivland utilised the space to draw together several of strands of her art-work, some of which had appeared in *Reproductions II* (at Domobaal in London in 2003, 2009 and 2013) and *Femmes Folles de Leur Corps* (at Galerie Bugdahn und Kaimer in Düsseldorf in 2013). By gathering (old and new) art-works together, *Folles de Leur Corps* offers an opportunity to think through some enduring themes that stretch across Kivland's work. My interest in her work lies in its weaving together, sometimes literally, of affect and politics through the body – through bodies of different kinds.

In particular, I am keen to stage a conversation between signature works in Kivland's *Folles de Leur Corps* exhibition and Rancière's arguments about the relationship between aesthetics and politics. This conversation, in some ways, is organic: each seem to be making a similar point, yet, for me, Kivland's art-works require us to think horizontally about co-existing bodily regimes. This is about more than who exactly is crazy about their bodies and about what makes them crazy. It is about the shifting meanings of bodies and being crazy. Explicitly, what is at stake is femininity, female sexuality and the female body, where the overt or

Bodies, Affects, Politics: The Clash of Bodily Regimes, First Edition. Steve Pile.
© 2021 Royal Geographical Society (with the Institute of British Geographers).
Published 2021 by John Wiley & Sons Ltd.

extreme expression of passion, especially sexual, is taken to be crazy: so, are passionate women deranged, enthusiastic, absurd or aggressive? Here, we can think back to both Emmy von N. and Dora (see Chapters 4 and 5).

Although Kivland's work challenges us to think about the production of femininity, she is well aware that bodies are structured by class relationships and by racialisation – and other bodily regimes. However, it is Kivland's interventions in what we can all the aesthetic unconscious that are of particular interest in this chapter. Bluntly, Rancière argues that social structures rely upon a distribution of the senses that assigns bodies to a proper place and thereby enables bodies to be policed. The distribution of the senses might also be taken to be, or to have, an aesthetic unconscious. This is the set of tacit understandings of what capacities bodies have and what they can do. Certain bodies can do some things, but not others; be in some places, but not others. We have seen this in Chapters 2 and 3. However, I have argued, the unconscious is more than a set of tacit assumptions, it is also a set of processes that repress, distribute and communicate traumatic and taboo contents. For me, Kivland seeks, along with Rancière, to make the assumed visible, by playing with aesthetic forms, making them available for conscious reflection. However, she moves further than Rancière by playing with the multiple and dynamic processes that produce aesthetic objects. Her objects are 'ugly beautiful' and 'beautifully ugly', making play of the contradictions, ambivalences, overdeterminations and indeterminacy that produces objects and subjects. And this is political for it interferes with the policing of the senses.

Folles de Leur Corps

Setting foot in the gallery on 5 October, visitors were greeted by a flurry of activity. Added to the 18 art-works, Kivland had arranged for music to be played and for eight young women to circulate amongst the guests, inviting them into conversation. The music creaked out of an old portable LP player, with the playlist being taken from the Stefan Szczelkun's Agit-Disco project (see Szczelkun and Iles 2012). In French, the melodies hinted at wartime Paris. Or perhaps revolutionary songs from earlier times. The air seemed to be punctuated with calls for an act of resistance. The young women, meanwhile, walked amongst people bare-footed, wearing linen dresses with sentences in red silk written in French across their chests. On their heads, a few wore plain red Phrygian-style caps. These caps were made famous during the French Revolution as a symbol of liberty. The association between liberty and the Phrygian cap goes back to ancient Rome, but it is based on a misconception. The Phrygian cap had become confused with the *pileus*, the conical brimless felt cap worn by emancipated slaves in ancient Rome. During the *manumissio* ceremony that set the slave free, a *preator* would touch a slave on the shoulder with a rod and pronounce the slave free, then a *pileus* was placed on top of the slave's shaven head as a symbol of the new freedom.

Fortunately, Kivland had not asked her volunteer women to shave their heads to wear their caps. Being barefoot was hardship enough in the name of art: the concrete floor was unforgivingly cold, and a good number of conversations started with observations about frozen toes.

Elsewhere in the exhibition, there was a screening of Peter Watkins' 5-hour and 45-minute epic *La Commune (Paris 1871)* (2000). The film uses a large, mainly nonprofessional cast, including many immigrants from North Africa, who conducted their own research for the project (often starting with little or no knowledge of the commune). The film re-enacts the Paris Commune as if modern TV news cameras were on the scene – forcing viewers to draw comparisons between the Paris Commune and more contemporary events. Alongside this, the exhibition accommodated a bookshop and reading room: amongst those on display (from Kivland's personal collection) and available for purchase were books by Rancière.

Kivland's desire to connect art and politics intersects, at various points, with Rancière's desire to understand the relationship between aesthetics and politics. This chapter is, in part, structured around those moments where their shared desires seem closest. However, Kivland also shows that bodily regimes are produced in a variety of different ways – and these render the body both mutable and indeterminate. Or, better, she reveals the limits of the work that goes into making bodies coherent and readable. Here, she shares a project with Rancière, as both wish to lay bare the means through which 'common sense' is produced, shared, maintained: that is, the ways that bodily regimes organise senses in common.

In *The Politics of Aesthetics*, Rancière argues that there is a 'factory of the sensible' (2004, p. 39). Through the plurality of human activities, the senses are woven together into a shared world of the senses. He argues that this shared world of the senses is never just the sedimentation of intertwined human activities. Rather, shared worlds of the sensible are produced by modes of production, which combine ways of living, working and being-in-place. These modes of production define the relationship between the ordinary (work) and the exceptional (art). For Rancière, art disrupts the ways that the distinction between the ordinary and the exceptional is produced and maintained. Further, art destabilises the separation between the idea and the sensible material that is assumed (only) to represent it: for an illuminating discussion, see Tolia-Kelly (2019).

Art brings to light, and calls into question, the distribution of the sensible and the distribution of domains of activities: especially, the distinction between those who act and those who are acted upon. This parallels Kivland's desire to both highlight and to disrupt the entwined social formulae 'men act, women appear' (Berger 1972, p. 47) and 'men look, women look at themselves'. To disarrange these formulae, for Kivland, requires more than catching people in the moment of looking and, then, asking that they reflect on the act of looking in that moment. She seeks to solicit a variety of different affects by disturbing the separation of looking and appearing, by inducing feelings from squirming discomfort to the frisson of unconscious fears and desires.

In the production of art, Rancière wishes to see a new distribution of the world of the senses, of the sensed world. This means turning over the relationship between manufacturing (making things by hand) and visibility (the ways that things are seen). Thus, for Rancière, art anticipates new ways of living and being in the world. Perhaps, for Kivland, this is a necessary idea, but insufficient. New possibilities for new worlds of the sensible are not just to be realised in thought, they are to be made, embodied and lived. This means inhabiting the body in new ways. It is no accident that her art-works include re-worked clothes. Her art-work is designed to actualise new possibilities for the senses: she wishes to detonate, to provoke, to cause to happen. This marks a different terrain for the relationship between art and politics. While Kivland (literally) fabricates new possibilities for the senses, Rancière is more circumspect. For Rancière, art cannot determine its relationship to politics. He asserts:

> The arts can only ever lend to projects of domination or emancipation what they are able to lend them, that is to say, quite simply, what they have in common with them: bodily positions and movements, functions of speech, the parcelling out of the visible and the invisible. (2004, p. 14)

The arts do not enjoy an autonomy where they can pre-determine their relationship to the political. Regimes of aesthetics can disrupt common sense worlds, which organise and assign bodily positions, which designate who is and is not heard, and which determine what can be seen and what cannot. Yet, Kivland sabotages the idea that aesthetic regimes are themselves coherent and integrated, rather focusing on the many ways that these fragment, proliferate and mutate bodies and identities. This lends a different relationship between art and politics, in which art is itself a disruptive of itself as a sensible regime. Rather than offering to re-organise the senses along different lines, the aesthetic regime of art reveals the 'real fiction' of bodies and politics, in part by revealing the plurality of corporeal regimes through which subjects organise their relationship to the world.

Modes of Production

Kivland describes the dresses in her art-work *Des Femmes et Leur Éducation* (2014) almost as if writing for a trade magazine (see Figure 7.1). They are made from a single template (size 38), each ready to be adjusted to the size of the woman who will wear it. Each seam is sown with a running stitch of red silk of the type called *soie de Paris*, a thread with six filaments which is smooth, soft and lustrous. That's not quite all. Kivland adds, the *soie de Paris* thread is 'difficult to work for it snags on rough fingers [. . .] To work with silk, one must take extra care of the smoothness of one's hands' (Kivland 2014, p. 40). The dresses have to be fabricated and this work is itself to be acknowledged (in contrast to the art that hides its own

Figure 7.1   *Folles de Leur Corps*, London, 2014. Handmade wool felt Phrygian bonnets, lined with silk, sewn maker's labels (from *Mes Sands-culottes*, 2014). Linen dresses sewn with *soie de Paris* (*Des Femmes et Leur Éducation*, 2014). Source: Steve Pile.

manufacturing process, so as to appear artfully unskilled). Thus, the work in the work of art (skilled or unskilled, smooth or rough) is also productive of regimes of the body.

The linen dresses in *Des Femmes et Leur Éducation* echo the calico dresses in another work in the exhibition. In *Pour Ma Vie Citadine* (2014), Kivland takes three photographs from the May 1968 edition of the fashion magazine *Mode Traveaux*. These are presented almost life-size and accompanied by three hung calico dresses (see Figure 7.2). Reminiscent of *Des Femmes et Leur Éducation*, each dress seam is stitched with *soie de Paris* red silk. The contrast between the material of the dresses and the silk thread in both works is significant: 'the calico of the fitting pattern is a coarse weave, unbleached and unprocessed, originally from India, and at times banned from export in favour of local industrial production. Under colonial rule, India was reduced to a supplied of raw material' (Kivland 2014, p. 40). For Kivland, understanding the mode of production of the work of art extends beyond the quality of the fabric and the nature of the labour process, the relations of production stretch geographically and historically – into other modes of exploitation.

The oldest producer of silk is China, but from the nineteenth century onwards the best quality silks were produced in Japan. Silk's qualities often connect it to

Figure 7.2   *Folles de Leur Corps*, London, 2014. *Pour Ma Vie Citadine* (For my city life), 2014. Calico dresses droop from three deer hoof coat hangers. Source: Steve Pile.

sexual desire. Indeed, in an unseemly expression of fetishism, the silkworm moth is often described as a 'machine devoted to sex'; in quote marks, despite being unattributed (and possibly erroneous). The Latin name for the mulberry silkworm is *Bombyx mori*. By contrast, then, *mori* connects the silkworm to death. The vast majority of silk is unwoven from the cocoons of mulberry silkworms, which are easier to unravel once the silkworm has died (often by boiling). According to PETA, about 6600 silkworms die to produce 1 kilogram of silk. Thus, the red thread leads a double life: on the one hand, connecting the dresses to sex and, on the other, to death.

On the dresses, the silk leads another double life. The red thread spells out sentences, hand-sewn in French; more specifically, hand-written in the style taught in French schools. Thus, in Figure 7.1 (showing *Des Femmes et Leur Éducation*), we can read what appear to be injunctions or aphorisms: *je me laisse aller sans me défendre* (I let myself go without defending myself) and *je suis pas Besoin de ma repaître d'illusions* (I do not need to feast on illusions). Kivland tells the women wearing the dresses that these phrases are drawn from a text by Laclos on the education of women. This is not the whole story, however. The source material for the phrases is, as Kivland states, Pierre Choderlos de Laclos' 1783 unpublished treatise *Des Femmes et De Leur Éducation*. What Kivland draws upon, in particular, is Laclos' description of, what he calls, 'natural women'.

In his text, Laclos distinguishes between uneducated, working class 'natural' women and educated, aristocratic 'social' women. While he values both kinds of women, Laclos emphasises the strength, character and beauty of natural women. Even so, Laclos sees education as a way of liberating women from slavery. Through her dresses, Kivland notes that it is only through a change in her condition that women can be freed; freed from the condition of being a possession. Yet, where Laclos writes *about* women, Kivland converts his sentences into the first person, changing 'she' into 'I'. The use of 'I' creates a sense of ownership and assertiveness of the phrases, not just for the women wearing her calico dresses, but also for women in general. Thus, Kivland refuses Laclos' division between 'natural' and 'social' women. Kivland's women are both natural and social. Their bodies are bare and written on, yet also clothed and speaking. They are to be looked at, but awkwardly, challenging people to talk to them, to listen to what they have to say. Their bodies, with their cold feet, cease to be objects, as they embody a demand to be paid attention to.

The red thread, woven into the linen and calico dresses, is a revolutionary thread: not a revolution that returns us to the same place, but an overturning of a natural order that demands that the proper place for women is that they are 'bodies', silent and decorative. Its doubled double life intended to disrupt any intention to confine its meaning to a single frame of interpretation. It is not just the red thread that leads a doubled double life, though, so too do the dresses. Thus, the use of linen and calico creates an ironic juxtaposition. Linen is, literally, used to make money. So-called paper money is traditionally made of a mix of cotton and linen fabrics: for example, US dollar bills are 25% linen. Conversely, calico is cheap.

The *Pour Ma Vie Citadine* dresses are made from calico, a plain-woven textile made from unbleached cotton that is often not fully processed. Originally, the fabric was created by weavers in the eleventh century in Calicut (Kozhikode), on the Malabar Coast in southwestern India. By the fifteenth century, calico from Gujarat was being exported to Egypt. In the seventeenth century, a trade in calico developed with Europe. Cheap calico printed fabrics, imported by the East India Company, became increasingly popular. By the eighteenth century, calico was popular enough to threaten English production of wool-based fabrics, such as worsted. The importation of dyed or printed calicoes from India, China and Persia was banned in 1700 to protect the English wool industry. However, the restriction on imports of finished cloths did not diminish demand. Even so, the restrictions on trade effectively reduced India to being a producer of raw materials. Calico, therefore, is an important marker of British imperialism and its power to transform economic relationships in its own favour: enabling the supply of cheap unprocessed cotton fabrics, protecting the domestic production of woollen fabrics, while at the same time deindustrialising Indian textile production. Calico is cheap, but at a price: a price paid, not in England, but by other people in other countries. Kivland's dresses, thus, invite parallels with contemporary sweatshop arrangements that are etched into global commodity production.

In the linen and calico dresses, Kivland juxtaposes – and seemingly contrasts – two different kinds of fabric: on the one hand, there is the silk thread that connotes sexuality and luxury; on the other hand, there is the linen or calico that denotes exploitation and economy. Yet, taken together with the Phrygian cap, there is a clear resemblance to the figure of Liberty in Eugène Delacroix's 1830 painting, *Liberty Leading the People*. In the painting, Liberty is bare-footed, wearing a plain yellow dress, with a red sash around her waist, and a red cap. Liberty is striding forwards, over a fallen barricade, encouraging the men behind her to action. Liberty herself is a dual figure, on the one hand a figure representing the spirit of freedom, while on the other hand she also a woman of the people. Thus, we might observe, Liberty is simultaneously an allegorical and a natural women (to use Laclos' term). As are Kivland's women. They both represent, and embody, the spirit of liberty. Both as Revolution, and also as a more prosaic refusal to settle into a single dominant frame of meaning.

Skin over Skin

The dresses and bare feet of the natural women in *Folles de Leur Corps* have a parallel in other works by Kivland that explore, more directly, the relationship between women's bodies, their skins and the ways that these have been written over. In particular, she explores the ways that women's bodies are concealed and revealed as a fetish. The fetish in this understanding is ironic. Kivland reminds us of Karl Kraus' observation that: 'There is no more unhappy being under the sun than a fetishist who pines for a boot and has to content himself with an entire woman' (cited by Apter 1991, p. 28). In her 2012 series, *Mes Lignes Parfait* (My perfect lines), this irony is developed by highlighting the way that lingerie both covers the body, but also gives the body a particular shape (Figure 7.3). By adding black ink and pink water colours to lingerie designs, the female body becomes visible through its absence. Kivland thereby highlights the way that lingerie (bras, girdles, corsetry) functions as 'second skin' (see Steele 2014; also Prosser 1998; and Cheng 2010). This second skin is, strangely, a layer between skin and the world; sometimes seemingly replacing skin itself, sometimes becoming the outermost layer of second skin. By substituting for skin, and becoming skin, lingerie functions as a fetish: that is, as an object with almost supernatural powers to solicit (sexual) responses. Ironically.

The designs themselves are taken from original 1950s women's magazines. One such design is by Jacques Fath. Fath was a popular designer in 1940s and 1950s Paris, along with contemporaries such as Christian Dior. Indeed, some of his young designers would go on to become famous in their own right, such as Hubert de Givenchy and Guy Laroche. Fath's clients included film stars such as Ava Gardner, Greta Garbo and Rita Hayworth. And, notoriously, Eva Perón. Designed to 'improve' posture, his girdles combined new materials with 'new

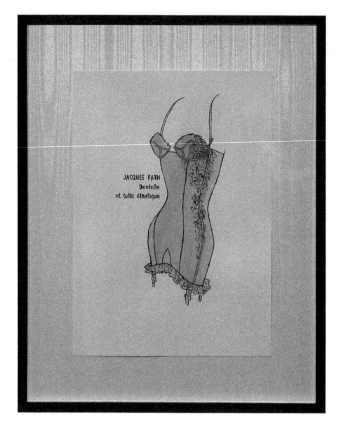

Figure 7.3   *Reproductions II*, 2013, London. *Mes Lignes Parfaits – Jacques Fath*, 2012. Ink and watercolour on Canson paper, mounted on ivory silk moiré paper. Source: Steve Pile.

lines' for the body. Lingerie acted as an intermediary between the 'outer look' of the body and the body itself. It ensured that the bodily regime of the 'outer look' was provided with an under-girding. The girdle ensured that the female body was not only provided with the latest shape, but also that the garment itself provided a skin for the woman's skin. Not simply an act of architecture, lingerie becomes an attribute of sexuality itself, with all its supernatural powers. Or, rather, lingerie provides an alternative model for the ego and its passions, one that can be shaped along new lines. Kivland intends to subvert the increasing detachment of the lingerie-as-fetish from the bodies that they give shape to. This is not simply to restore the absent woman's body, but to laminate the skin of lingerie to bare skin. Thus, lingerie-as-skin represents a protection against the threat to disavow, or even excise, female sexuality and desire.

This approach accords, in some ways, with Sigmund Freud's discussion of the fetish (1927), as Kivland is interested in the ways that substitutes for, or supplements to, the body can become sexualised and, indeed, become more sexual than the body itself. Yet, Kivland utilises Karl Kraus' joke about the unhappy fetishist (who yearns for a boot, but has to be content with the entire woman) to democratise the sex of the fetish. This runs counter to Freud's account of the fetish. Freud's assumption is that sexual fetishism is exclusively male and grounded in male infantile fantasies about women's bodies and fears about their own bodies: especially, the bit they hold dearest. Nonetheless, significantly, there is a certain hesitancy in Freud's analysis of fetishism:

> If I now state that a fetish is a penis substitute, this will no doubt come as a disappointment. I hasten to add, then, that it is a substitute not just for any penis, but for a specific and very special one, one which is of great significance in early childhood but which is subsequently lost. That is to say, it should normally be renounced, but it is precisely the purpose of a fetish to prevent this loss from occurring. To put it more plainly, a fetish is a substitute for the women's (mother's) phallus, which the little boy once believed in and which – for reasons well known to us – he does not wish to give up. (pp. 95–96)

Freud is aware that it will be disappointing to learn that fetishism lacks psychological ingenuity, it is simply a way to deny the threat of castration. What the fetish does is conceal the woman's lack of penis and therefore protects the fetishist from the traumatic idea of castration. The fetish itself, less disappointingly perhaps, therefore preserves the idea that women have – or have had – the phallus. To be castrated, women must have had a penis. The fetishist, albeit unconsciously, believes in the female penis (or, the phallic woman). But this idea has to be denied because it confirms the threat of castration. As a result, the fetishist finds a female penis substitute, a substitute that is already connected to the idea of the female penis. The choice of fetish, then, is often based on nearness to female sex, so feet and footwear. And, of course, lingerie, whether hard (in the form of corsets and girdles) or soft (in the form of stockings and underwear). Yet, Freud's account does not quite stack up even for Freud.

For Freud, the horror of castration lies at the heart of male sexuality. Yet, the fetish throws a distinct spanner in the Oedipal works. He observes:

> Admittedly, we cannot explain why some men become homosexual as a result of this experience [the horror of castration], others ward it off by creating a fetish, while the vast majority overcome it. (p. 97)

Given that the horror of castration is the same for all men, it provides no clue as to why men respond differently to it. This problem is compounded when Freud discusses particular fetishes. He argues that fetishes act as blocks on memories that have traumatic content. The fetish, then, is a historical marker of the moment when the boy's curiosity about women's bodies discovers the horror of castration:

> Thus feet or shoes owe their prominence as fetishes, at least in part, to the fact that the curious boy looked at women's genitals from below, from the legs up; fur and velvet are – as we have long suspected – fixations on the sight of pubic hair, which should have been followed by the longed-for sight of the female member; pieces of underwear, so commonly adopted as fetishes, capture the moment of undressing, the last point at which the woman could be regarded as phallic. (pp. 97–98)

The fetish has a dual function: it affirms that women have been castrated, yet acts to deny this castration. Thus, women's genitals become uncanny: threatening to panic the (unhappy) fetishist with the horror of castration. The fetish, therefore, forestalls uncanny experiences of actual women's bodies by acting as a screen onto which men can project their sexuality. One example Freud gives is of the girdle, of the kind that can be worn as a swimming costume, of the kind that Kivland uses in *Mes Lignes Parfait*:

> This piece of clothing completely concealed the genitals and the difference between them. According to analysis [of an unnamed patient], it signified both that women were castrated and that women were not, and furthermore, it allowed for the assumption of male castration, because all these possibilities could equally well be hidden beneath the girdle. (p. 99)

The fetish, therefore, fuses incompatible beliefs: that women are castrated, that women are not castrated. Yet, Freud elaborates. The fetish can also fuse other incompatible beliefs: that women still have a penis, that women are castrated by their fathers. Freud's somewhat disappointing statement that fetishes are, simply, substitutes for the penis, by the end of his paper, is unravelling. The fetish is a composite of contradictory ideas; its sexual force, the product of the relationship between men's and women's bodies and ideas about men's and women's sex; it is ultimately indeterminate, the choice of fetish and of fetishism appears arbitrary, inexplicable.

So, for Freud, the fetish is a sign for male horror of the female body and of male anxieties about their own bodies. The traumatic affect is repressed, producing an indelible aversion towards actual female bodies for fetishists. Yet, Freud's inability to explain different responses to women's bodies, as well as the seeming arbitrariness of his examples, seems to suggest that he may be right that the fetish actually opens up all possibilities: that is, that the fetish creates more possibilities than Freud can acknowledge. There is, then, an opportunity to re-imagine the fetish, to provide it with other pathologies – and to replace Freud's horror of women's genitals with something else, something like the frisson of knowing and not knowing, of seeing and not seeing, of touching and not touching.

In Kivland's work, this requires an exploration of the way the fetish fuses contradictory ideas. Lingerie, then, contains opposing ideas about sexuality, skin and the body: one the one hand, it represents the surface of the body, yet on the other hand it obscures and shapes the body. However, rather than simply laminating opposing ideas about castration, rather than affirming the repression of affects of horror, Kivland highlights the significance of absence and presence – of providing a form for the passions – in the formation of sexuality and bodies. Kivland explores this relationship in other works, too. For example, in *Mes Vedettes* (My stars) (2012) takes postcards of French female film stars (such as Eva Marie Saint) and paints red onto their already lip-sticked mouths and white into their teeth and eyes. By over-emphasising the nature of the look, the function of these images as 'sex symbols' is called into question. The women 'over appear'. They look too much. A similar effect is achieved in Kivland's *Mes Plus Belles* (My most beautiful) (2010). Using 1968 women's magazines as a source, Kivland selects images of women that are looking directly into the camera. These images are painted over, ensuring that the women have blonde hair, dark eyes and pink lips. The effect is to exaggerate the look of the women, even while the exaggerated colours interfere with the way the women look. In these works, by exaggerating the way the women look – are made to look – Kivland reveals how the women themselves disappear in the process of the production of images of women.

The choice of 1968 women's magazines in *Mes Plus Belles* is, of course, a deliberate reference to the political events occurring in Paris at the same time. The reference is ironic: the idea of women as political agents also disappears in women's magazines. As a counter-action, Kivland designs clothing for women with explicit political messages. In *Mes Devices* (My devices) (2008), Kivland embroiders, on the inner arm of vintage kidskin 'matinee' gloves, the words *liberté, égalité, fraternité, ou la mort* (see Figure 7.4).

In her description of the work, Kivland explains:

> The motto of the French Revolution was painted on the walls of Paris. Revived in 1830, with the short-lived revolution of that year, it was not adopted officially until 1848, the year of revolution that came too soon, and the consequence of death disappeared, along with other earlier formations. (2014, p. 63)

Kivland seems to restore the original motto of the French Revolution, yet also calls into question the history of the motto at the same time. This raises more than the cleansing of death from the history of the French Revolution. It cleanses revolution of its visceral anger. It removes the stake of revolution: death. This converts the French Revolution into a desire for a holy trinity of ideals: liberty, equality, brotherhood. Yet, *ou la mort* is a sharp reminder that revolutions are not simply a response to inequality, injustice and social division, but also to the threat

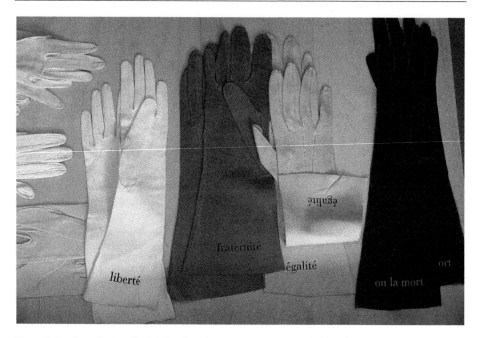

Figure 7.4   *Reproductions II*, 2013, London. *Mes Devices*, 2008. Antique kidskin gloves, foil letterpress blocking. Source: Steve Pile.

to life itself. The stake of life itself is juxtaposed, in *Mes Devices*, with symbols of luxury and aristocracy: kidskin opera gloves. Kidskin does double duty, of course, connoting both the death of the young goat from which the gloves are made, but also the skin of human children. The matinee-length gloves became popular in the 1950s, as part of the formal attire for women at afternoon tea, parties and weddings. The matinee glove was one amongst an armoury of gloves: there were day gloves, driving gloves, and opera gloves. They were made of leather, cotton, nylon, wool. Many were embroidered. These skins for the hand covered skin itself. They both hid skin, but also made the hands more visible: they were a declaration of femininity and style. By adding *ou la mort*, Kivland is commenting on 1950s women's fashion, but also the stake of fashion – with its potential to create a form of social death. The intimacy of this arrangement is hinted at by embroidering the French revolutionary slogans onto the inner arm. They are there to remind the wearer of the stake of fashion – that its stake is life itself. Except, it is not. The words, rather, call up an older political imperative. Whilst fashions – and political slogans – come and go, and are infinitely revisable (and cleansable), the fury of revolution remains an ever-present possibility.

In other works, Kivland is determined to confront her audiences with this relationship between death and revolution – through the bodies of beasts.

The Secret of Fetishism

Amongst the stuffed animals in *Mes Bêtes Sauvages* (see Figure 7.4), there are foxes carrying books by Karl Marx. One can be found carrying a 1969 French edition of Volume 1 of *Le Capital*; another, a 1945 edition of *Salaires, Prix et Profits*; yet another, a 1994 edition of *Wage-Labour and Capital; Value, Price and Profit*. Perhaps these editions, taken from different decades attest to the enduring interest in the writings of Karl Marx. Foxes have a reputation for cunning and being sly. One sly fox, its head half emerged from under a cotton dress, is carrying a 1949 edition of *Trois Essais sur la Theorie de la Sexualité* by Sigmund Freud. Their appearance in the exhibition draws inspiration from a footnote in Chapter 2 of Marx's *Capital, Volume 1*.

We can reconstruct Marx's train of thought by inserting the footnote into the main text. It would read this way:

> Commodities are things, and therefore lack the power to resist man. If they are unwilling, he can use force; in other words, he can take possession of them. [Footnote 1] In the twelfth century, so renowned for its piety, very delicate things often appear amongst these commodities. Thus a French poet of the period enumerates among the commodities to be found in the fair of Lendit, alongside clothing, shoes, leather, implements of cultivation, skins, etc., also *femmes folles de leur corps* (wanton women). (p. 178)

According to the translator, Ben Fowkes, the list of commodities Marx refers to is drawn from Guillot de Paris' 168 verse poem *Le Dit du Lendit* (1290). The religious festival of Lendit began with the opening of the reliquary of Saint Denis on 9 June 1053. Also, on public display were relics of the Passion: a nail from the cross and part of the crown of thorns. As the religious festival drew ever more pilgrims, so the trade in commodities grew around the festival. By the early twelfth century, a large fair had developed in the plain between Saint-Denis and Paris, which lasted from the second Wednesday in June until 23 June. The festival drew merchants from dozens of towns (including towns in England and Belgium), mostly specialising in cloth production, including silks. The fair was also significant for its livestock and leather markets. Guillot's poem starts by saying it is written in honour both of the merchants (from barbers to tanners, from goldsmiths to bankers) but also of the 'world's most royal fair' (*la plus roial Foire du Monde*). The poem catalogues the merchants and commodities on sale. The commodities he lists include spices, cheese, skins, shoes, jewellery, cows, sheep, horses, salt, bread, candles, but only to beautiful ladies (no wanton women).

Rather, Marx seems to have been referring to a contemporaneous poem by Guillot, *Le Dit des Rues de Paris* (1280–1300), in which he refers to *dames pour louier* (ladies for rent). It is Marx, then, who converts *femmes a louer* (women for rent) into *femmes folles de leur corps* (wanton women). This alters the relationship between women and their bodies: Marx converts women from owner-renters into commodified bodies. In Marx's terms, the women are alienated from their own bodies, as relationships between people become subverted into relationships between things (Figure 7.4). Between Guillot and Marx, we can see that women occupy a paradoxical place, for they both possess the commodity and are the commodity. Contra Marx, women – as a commodity – have the power to resist men. For Kivland, it is not only women's bodies that offer the potential to resist men. Animals, also, raise the question of who owns the body and the relationships through which their bodies are exchanged (as commodities).

Thus, Kivland draws particular attention to Chapter 1 of *Le Capital*, which is titled *La Marchandise* (The Commodity) (1867). At the outset, Marx talks about the ways that labour transforms nature into commodities. In doing so, labour transforms itself. Kivland herself is interested in moments of transformation, both of material forms and bodies, but also of social relations. For her, these are the moments when aesthetics and politics are lived and experienced as entangled. Her interest, then, is the way that women mediate between aesthetics and politics. Transformation is not just economic, it is also about affects. The shift between *femme a louer* and *femmes folles de leur corps* not only marks the passing of women as rational property owners, but the installation of women's desire and women's relationship to their bodies as a craziness. This craziness connects two distinct trains of thought. On the one hand, there is a bond between the commodity, the body and female sexuality. On the other hand, the craziness associates ideas about hysteria, the body and female sexuality. Between the commodity and hysteria, arguably, is the fetish – a male hysterical response to the sight/site of the female body. Yet, Marx also has things to say about commodity fetishism, not coincidentally in Chapter 1 of *Capital*.

In the concluding section of Chapter 1, 'The Fetishism of the Commodity and Its Secret', Marx argues:

> A commodity appears at first sight an extremely obvious, trivial thing. But its analysis brings out that it is a very strange thing, abounding in metaphysical subtleties and theological niceties. (p. 163)

The source of these mysteries lies in the way that nature is transformed, by labour, into a commodity. Wood becomes a table. The table is still wood, but as a commodity it becomes both sensuous and an idea:

> it not only stands with its feet on the ground but, in relation to all other commodities, it stands on its head, and evolves out of its wooden brain grotesque ideas, far more wonderful than if it were to begin dancing of its own free will. (pp. 163–164)

The dancing table is a reference to the spread of spiritualism across Germany from the mid-1850s onwards. Marx's metaphor indicates that although commodities, such as the table, are made of dull matter, they become mystical and enigmatic.

He argues that the mystical power of commodities lies in their social relations: commodities are not just a physical form, they are also a social form. Indeed, the social form of the commodity can become detached from the physical form and the material relations that constituted it. Thus, the value of a commodity is less tied to the labour that went into producing it than to its social (exchange) value. Social value substitutes for labour value. Indeed, it is through the superimposition of social value onto labour value that the thing becomes a commodity. Thus, Marx argues, the sensuous physical form becomes supra-sensible.

Indeed, Marx was well aware that commodity fetishism was supra-sensible because the commodity is connected not just to needs, but to desires and wants. In his discussion of hoarding, he observes: 'The hoarder [of gold money] sacrifices the lust of his flesh to the fetish of gold' (p. 231). Commodity fetishism and sexual fetish are connected, at their root, to the disavowal of the flesh and its replacement with another object of desire. Kivland shows that the distribution of the sensible is correlated with the distribution of desire – who and what counts as a desiring subject, who and what counts as an object of desire.

Disavowal is integral to the supra-sensible quality of the commodity. It is disavowal that enables the commodity to appear to have an autonomous life, situated in fantastical, rather than material, relationships. As a result, the commodity can appear to have its own life and meaning, independent of the labour that produced it. This is perhaps why Kivland stages a scene with a cobra and a stoat (Figure 7.5). At first glance, they appear to be fighting over a French edition of Marx's *Le Capital*. On the other hand, the stoat might be gifting or simply showing the book to the cobra. Stoats and cobras. What becomes clear is that the stuffed animals are awkward commodities, seemingly full of life, yet frozen forever in the moment of action. They conceal and display their death.

The calico dresses of *Pour Ma Vie Citadine* hung on deer hooves (Figure 7.2), the stuffed animals with red bonnets, the books by Marx and Freud are all commodities, all seeming to have lives of their own. They are, in Kivland's terms, charming. They lead charmed lives, for they keep secret the labour – the material relationships – that went into making them. This, then, is the essence of the fetish of the commodity. It is supra-sensibly mystical and affectively powerful, precisely because it conceals its ordinariness and its materiality. Like the sexual fetish, the commodity fetish contains opposing ideas: it is both life and death, both charming and dull; it can be glamorous and sexy, yet often also the product of pains-taking, ill-acknowledged hard work. Indeed, Kivland also calls into question how labouring is often seen as unskilled, revealing instead that the production of seemingly simple commodities (such as a calico dress) requires both skill and understanding of the nature of the material. On the other hand, the stuffed animals are given hand crafted politically resonant bonnets, thereby holding two contradictory ideas together: animals existing outside human culture and politics, while also existing within culture and politics – especially stuffed ones (see also Straughan 2015). Unfolding these layerings de-naturalises the seeming rawness of the commodity.

Figure 7.5   *Folles de Leur Corps*, London, 2014. The exhibition contains many taxidermied animals, in different works. In *Je Suis Une Petite Charmeuse* (I am a little charmer), 2013 and 2014, there are three stoats, wearing red wool Phrygian caps and fourteen squirrels, some with antique collars with contemporary inscriptions. *Mes Bêtes Sauvages* (My wild beasts), 2014, has eight squirrels, eight stoats and five foxes. The squirrels wear velvet ribbons, while the Jacobin stoats wear red Phrygian hats, while the foxes carry books by Marx, including a French edition of *Capital, Volume 1*. Source: Steve Pile.

And, we can add, shows that the commodity can happily embody both the supra-sensible and the super-natural.

To align this with Rancière, commodification creates sensible worlds. Thus, the commodity is a supra-sensible boundary marker in the distribution of the sensible. However, the commodity is also capable of destabilising that boundary and reorganising the distribution of the sensible: as, for example, when tables dance. By drawing attention to the commodity form, Kivland seeks to illuminate the commodity's role as marker for the margins of the sensible and as a site where the distribution of the sensible can be contested.

A Woman Who Doesn't Wear Perfume Has No Future

Fashion on (and of) the streets might, possibly, become incendiary. From the bedroom to the barricades, fashion may – like aesthetics – become part of a system of democratic principles that reinforce the desire for political transformation, revolution. At stake, in this transformation, is the woman's body, as a place where the distribution of the sensible is revealed and concealed. In Kivland's work, *Mes Petites Explosions* (2011–2012), the liquid contained in the perfume bottle, which is presumed to conceal the scent of the body, is revealed as explosive (Figure 7.6). In this work, Kivland speculates, *Allure* bottles might be used to make Molotov cocktails. And, we can add, it may be silk clad bodies that throw them. Kivland explains:

> Scent bottles (*Allure* by Chanel) are transformed into Molotov cocktails [. . .] *Allure* [. . .] is a *modern* scent, designed by its nose, Jacques Polge, in 1996, and so there is a certain instability and historical inaccuracy (as in so much of the artist's work). It is a scent that has been reinterpreted extensively in skin creams, for it is associated with a creamy effect, soft on the skin [. . .] in soaps that leave the skin clean, smooth, and supple, enhancing its beauty, leaving it like silk; the advertising declares its purity. (2014, p. 52)

In *Mes Petit Explosions*, women's bodies do not appear. Instead, they are evoked through their smell. Yet, the scent of women is about more than smell. It connotes other qualities of the feminine, such as softness, suppleness, cleanliness. Women's skin, thanks to *Allure*, becomes like silk. Jacques Polge's nose becomes the measuring device not simply of how women are to smell modern and alluring, his nose also evokes women's cleanliness and creaminess. As Juliet Flower Maccannell points out in her commentary on Kivland's piece (2014, p. 53), Freud placed the devaluation of the olfactory senses at the heart of 'civilisation'.

Figure 7.6   *Folles de Leur Corps*, London, 2014. *Mes Petite Explosions* (2011–2012). Allure scent bottles on a polished steel shelf with pink wedding tulle. Source: Steve Pile.

Freud suggests that, in the process of civilisation, the body and the senses are reorganised. Strenuous efforts are taken to make the human body seem less (and less) animal: genitals, smells, menstrual cycles are all increasingly concealed.

Sexuality, in particular, is reorganised through the imposition of family norms and gender differentiation – and through taboos, which smuggle in a degree of repugnance and disgust with sex itself. Thus, through the civilising process, the body becomes a target for repression. In a lengthy footnote to Section IV of *Civilisation and Its Discontents*, Freud suggests that the devaluation of smell is central to the repression of the body, as it is critical in the denial both of sexual and of excretory functions (1930, p. 43). There are parallels here with the idea of the fetish: the body is both a source of disgust, yet also desire. The body is consequently reorganised: natural smells are acknowledged so as to be disguised. Covering the body's smells enables the body to be recognised as civilised, but especially for women. Coco Chanel knew as much. 'A woman who doesn't wear perfume has no future', she observed.

Jacques Polge's nose is the means through which the woman's body becomes both mysteriously and powerfully attractive (alluring), and also disguised and civilised. *Allure* has as its target the natural woman: this is not simply an erasure of the body, but rather an organisation of both the body and the senses. It makes the woman available for sex by ensuring that her body is encountered as civilised – *modern* – to smell, to touch and to look at. *Allure*, then, draws on a distinction, much as Laclos did, between the natural and the social woman. In *Allure*'s case, the regime of the senses is organised around scent, but also envelops touch and the visual. What Kivland is seeking to do is reorganise the mystery and power of the woman's body, by scenting it otherwise. Civilisation has its discontents.

Kivland imagines that the *Allure* bottles can be filled with petrol, stuffed with expensive wedding tulle as a wick, and used as a convenient, charming, improvised incendiary device. The Molotov Cocktail simply requires a breakable glass bottle, a flammable liquid and a wick. The liquid is usually petrol (or similar). The wick is often soaked in kerosene, and held in place by bottle stopper. It is imaginable that the fire bomber could use alcohol-based perfumes. Although improvised incendiary devices like this were used in the Spanish Civil War, it was the Finnish name Molotov Cocktail that stuck during the second world war. Kivland manoeuvres the Molotov Cocktail into the history of the Paris Commune, thereby rebranding the fire bomb as both sexually alluring and politically incendiary. In a description of *Mes Pétroleuses* (2008), Kivland says:

> The term is pejorative, meaning both a cock-teaser, a woman who seeks to enflame a man's desire without satisfying it, and a woman who gives voice too vehemently to her political opinions. What a naughty tease! It is also the name for the women of the Paris Commune of 1871, accused of burning down much of Paris with petrol bombs, precursors of the Molotov Cocktail so popular during the *événements* of May 1968. (2014, p. 48)

In her blog about the art history of *La Pétroleuse* (12 June 2017), Lynn Clement discusses an 1871 caricature, attributed to Bernard. In the caricature, Bernard depicts as a deranged, middle-aged woman, wearing a red Phrygian bonnet and a red dress with an exposed breast. In her left hand, she is pouring *pétrole* from a watering can. With her left hand, she uses a lighted torch to ignite the petrol. Behind the woman, Paris burns. Next to her lies a dead soldier. At her feet are torn up cobble stones, used by the communards to create barricades, stained with blood. According to Lyn Clement, the *pétroleuse* is an echo of Delacroix's Marianne – a distorted and dreadful echo. Where Marianne represents the ideals of the French Revolution, the *pétroleuse* embodies anarchy, pouring fire onto democratic institutions, such as the *L'Hôtel-de-Ville*. In the aftermath of the Paris Commune, the figure of the *pétroleuse* represented both the demand for women's political rights and also its corruption into uncontrollable violence and destruction. Once again, the woman's body is a product of contradictory corporeal regimes: she is moral and political; she is destructive and deranged.

Kivland's charming little fire bombs await their incendiary moment. They contain wedding tulle, but not perfume nor petrol. Tulle is a lightweight, very fine form of netting. Its structure makes it stiff, strong and durable. Although mostly made of polyester, it can be made of silk. Tulle can be used to wrap gifts, for example at weddings, but it can also be used to add texture and shape. Kivland's tulle seems to be about more than providing her fire bombs with a wick: surely, they would burn too quickly? They are almost ethereal, as if a spirit or genie were escaping the bottle. If *Allure* and the pétroleuse position the woman's body between nature and civilisation, between arrangement and derangement, the wedding tulle seems unable to be contained within these dualisms. It suggests the woman's body escapes its casting within even opposing frames of meaning. Thus, Kivland seeks to highlight the contradictory regimes within which women's experiences of their bodies are enframed, yet also to enjoy and subvert those regimes. And to escape them.

Conclusion: Redistributing the Senses

In *Des Femmes et Leur Éducation*, Kivland draws upon Laclos' distinction between 'natural' and 'social' women. In distinguishing between natural and social women, Lalcos effectively creates two sensible worlds for women. Women are seen as having particular strengths, which revolve around judgments of beauty and refinement. Their bodies are packaged and re-presented. In each world, women have particular value – but there is also a cost. Women cannot inhabit both worlds. But, more importantly, in either world, women are confined by the sensibilities that seem to define them. In re-stating Laclos' 'they' (women) as an 'I', Kivland gives agency back to women. Yet, this is also paradoxical. For, this also shows how that 'I' is positioned within distinctions that seemingly define them. By using 'I',

Kivland then asks women to think about whether this 'I' is what 'I' think or do. Thus, Kivland not only draws attention to these worlds, she exposes their limits by asking women to reflect on whether this 'I' is their 'I'. Yet, of course, the 'I' belongs to no one. Orders of the sensible co-exist: the natural and the social; women and men. Distinctions constantly drawn and redrawn; more and more distinctions, coexisting. Perhaps, this is what makes people crazy about their bodies.

As the women wearing the linen dresses passed among the visitors to Kivland's *Folles de Leur Corps* exhibition, they engaged people in conversation about the artwork, but also chatted about whatever. When Kivland gathered them together to thank them for participating in the opening of the exhibition, they laughed and danced to the agit-music. Yet, their presence was also unsettling, deliberately. On a cold day, walking with bare feet on unforgiving concrete, it was impossible not to think about their discomfort. But, more acutely, reading the hand-sewn messages forced closer attention to their chests than is socially comfortable. Deliberately so. The women's presence in the exhibition resonating with its broader themes. Thus, the women enacted the boundary between comfort and discomfort: bodily, sexual, politically and socially. These boundaries are not quite in the same place. Kivland's women inhabit more than one bodily regime, both comfortably and unsettlingly. This crystallises the play of bodily regimes – of politics and affects – in the works of art. Kivland disconnects and reconnects bodily regimes, placing them side by side to beg questions about how these regimes operate through the body: that is, not simply as something that operates upon bodies, but that actively produces the body. This is perhaps clearest in her use of the bodies of animals to literally support other kinds of work: dresses, books by Marx and Freud, revolutionary caps.

Kivland explores the overlapping edges of different bodily regimes, but especially those between the body, sexuality and politics. Just because the edges overlap, this does not mean that they are integrated or coherent. Indeed, mostly, they are not. Edges are also gaps. In particular, there is a gap between women's experience of their own sexuality and the political means to effect change. This is why Kivland makes a petrol bomb out of a perfume bottle. She converts the horror of the Commune's fire-starting women into a force for liberation. But, this is not liberation from perfume, or from silk, or from sex, but a political demand for something else. This places Kivland and Rancière on the same page, their insistence is on exploring the effects of regimes of the sensible on subjects so as to be able to do them differently. What Kivland enables, however, is the opportunity for exploring political possibilities through history and experience. The critical moments to which she returns are, as importantly, all about Paris: especially, its Commune and 1968. The edges of bodily regimes are produced in, lived through, Paris. They are not, therefore, universals, but about the particular. People, women, may be crazy about their bodies, but that craziness is product of the historical materialism and ideologies of a particular space and place. In France. This desire to explore the political possibilities of history and of place is shared by Rancière, who has sought to track the energy of working class struggles in France (2012). This energy is not simply found in wild revolutionary moments, such as the Paris Commune and 1968, but in the ways people ordinarily express themselves. This struggle is partly about seeking a way to free themselves from the constrictions of labour and poverty. Here, the idea that there might be another way to live life is opened up by the things people become passionate, or crazy, about – and Kivland seeks to prompt these openings.

In this context, being crazy about the body is saturated in meaning. On the one hand, we learn from Marx that it is a euphemism for wantonness. This itself is ambiguous, potentially meaning female licentiousness or prostitution or both. However, it could also refer to hysteria, with women (and men) enacting their distress through the body. So, craziness circulates about the body. But, being crazy about the body also highlights people's narcissism and self-loathing. And, further, people's relationship to the bodies of others: other people's bodies can be objects of fascination and horror. As can people's feelings about other bodies, such as those of animals – and the dead. Craziness does not simply evoke the co-existence of multiple bodily regimes. Pointing this out would simply produce an ever-lengthening analytical shopping list of bodily regimes. Craziness tells us what the stake of living with these regimes is: it isn't simply that 'I' live in and with many regimes, but that these regimes rub up against each other and this has effects on the subject. It is these effects that have to be unstitched – and restitched. This requires attention to detail. This is why Kivland works and reworks the fetish: to imagine the body, sexuality and politics differently.

In her work, Kivland exploits the contradictory regimes that seek to circumscribe the body to explore the effects of ideology on the subject. For her, silk and scent are as much a part of a revolution of the body as calico and petrol. The woman stands for justice, but not an easy justice that is grounded in one version of femininity and sexuality. The body is, rather, mischievously indeterminate. Deliberately twisting histories, highlighting and hiding parts of the body, producing crafted and maladroit products, Kivland mixes and matches political imperatives for a less certain future. One in which perfume burns with democratic zeal, whilst also producing a silky beauty.

Chapter Eight
Conclusion: Bodies, Affects, Politics

Geographies of the Body and the Search for the Political

In the 1990s, the then 'new' cultural geography, inspired by its engagement with feminism, postcolonial and queer theory, and cultural criticism, generated a series of edited collections with the body as a newly emergent theme, amongst them are *Mapping Desire* (1995, edited by David Bell and Gill Valentine), *Bodyspace* (1996, edited by Nancy Duncan), *Places through the Body* (1998, edited by Heidi Nast and myself), and *Mind and Body Spaces* (1999, edited by Ruth Butler and Hester Parr). This work in cultural geography was capstoned by a sole authored monograph written by Robyn Longhurst, simply titled *Bodies* (2001a). Taken together, these works exemplified a broad-ranged critique levelled at, on the one hand, a lack of understanding of the power relations that structure people's bodily experiences and, on the other hand, the covert conceptual frameworks that structure understandings of the body, especially the tendency to dichotomise the body into binary categories such as male/female, masculine/feminine, straight/queer, white/black, able bodied/disabled and the like. Since the 1990s, the body has become a familiar theme within human geography. Indeed, uses of terms such as corporeality and embodiment have become common. Has this led to a greater engagement with the politics of bodily life? In two fairly recent essays, this question is asked explicitly by Robyn Longhurst and Lynda Johnston (2014) and by Kirsten Simonsen (2012). Their answers to this question are instructive.

Bodies, Affects, Politics: The Clash of Bodily Regimes, First Edition. Steve Pile.
© 2021 Royal Geographical Society (with the Institute of British Geographers).
Published 2021 by John Wiley & Sons Ltd.

In a meticulously researched paper, Robyn Longhurst and Lynda Johnston review papers about the body in the feminist journal *Gender, Place and Culture*, as part of its twenty-first birthday celebrations. They track the emergence of papers about the body from the mid-1990s onwards. The aim of this early work, as they put it, was to 'disrupt hegemonic masculinist structures of knowledge production' (2014, p. 268). And not just masculinist structures of knowledge. The concept of intersectionality was quickly introduced to enable other social inequalities, especially those structured by race, to be incorporated (for a review, see Valentine 2007; see also Esson et al. 2017). Longhurst and Johnston note the gathering quantity of this work over the lifetime of the journal. And not just in *Gender, Place and Culture*. This work has gained pace, and a place, across the bandwidth of journals publishing work by human geographers.

Helpfully, Longhurst and Johnston identify three major themes of work on the body since the turn of the century. First, there has been a focus on maternal bodies, inspired by Longhurst's own work on pregnancy (2001b). Second, there is an increasing engagement with (as they phrase it) the geopolitical body. This work explores the role of the body as it emerges in the contact zone between the so-called West and the Rest, including labour practices, bodies in conflict zones, and in the practices of development and aid. Finally, there is research into people's bodily experiences. This includes explorations of issues such as access to, and use of, public spaces (including shopping malls, toilets, beaches, bars and the like) as well as challenging assumptions about people with visibly different bodies. This has included a strong engagement with the 'trans' community (see Browne, Nash and Hines 2010), where common sense, normative categories for (sexing) the body become problematic.

Longhurst and Johnston see gaps in the literature, especially around the role of the body in research and around engagements with critical race studies and postcolonial theory, but it is hard to miss their uneasiness with the current direction of work on the body. Bluntly, while they celebrate the increasing quantity of work on the body, they worry that the body simply becomes 'little more than a ubiquitous marker of identity and difference, emptied of its power to unsettle the masculinist epistemology of the discipline of geography' (2014, p. 273). There are advantages, as they see it, to being able to see the ways that bodies are marked and experienced differently. However, they also see a problem:

While this might be useful in unfixing categories of individual or even intersecting subjectivities, this strategy might also carry with it risks, namely that the unmarked body may potentially elide the materiality and specificity of bodily difference. In the process, particular bodies – white, male, able-bodied, materially well off, Western bodies – come to be assumed and privileged over other bodies. (p. 273)

Longhurst and Johnston are carefully walking a tight-rope, here. On the one hand, they wish to avoid falling into the trap of seeing the body prior to its social construction. On the other hand, they also wish to acknowledge the physicality and materiality of the body, which may itself structure the social. The body teeters between its sociality and its materiality. This instability is to be welcomed, for it is here that we might be able to acknowledge the ways that some bodies become privileged over others. The politics of the body, then, emerges from the (masculinist, colonialist) epistemologies that underpin, and determine, the relationship between the body's materiality and sociality. The central problem with seeing the body as a site of identity and difference, then, is that this can fail to acknowledge the ways that the body is produced as a location within grids of meaning and power.

In some ways, Kirsten Simonsen maps out a similar problem in her quest for a critical phenomenology (2012; see also Simonsen and Koefoed 2020, ch. 1). For her, phenomenology has, since the humanistic geography of the 1960s, excelled in exploring people's bodily experiences. However, it has been less engaged around the political and around the significance of human agency. In her review of phenomenologies of the body, she offers a re-reading of Merleau-Ponty's work to establish the grounds upon which a politicised new humanism might be built. Chiming with Longhurst and Johnston's worry about the unmarked body, central to this politicised reconceptualisation of phenomenology is people's experiences of their bodies. Simonsen's proposal is to develop a politicised understanding of the body as a marker of identity and difference. Her approach is to recast experience around ideas of practice (drawn from Henri Lefebvre; see also Simonsen 2005). This enables Simonsen to (re)locate experience in its social contexts. This, then, allows critical phenomenology to explore people's experiences of the body as a marker of identity and difference in the context of unequal and injurious social power relationships. In this understanding, politics inheres in people's different experiences of the body.

Drawing on Lefebvre (1974, p. 170), Simonsen emphasises how the lived body is both productive of space and produces itself in space. This production is both social and material, practised and performed, in coexistence with others (following Merleau-Ponty). Seeing the body as producing and produced by a variety of lived spaces – from the home to transnational connections – begs questions, for Simonsen, about 'how this lived space involves participation, conflict and the appropriation of space for creative, generative, bodily practices'. Answers to these questions can be given a critical edge by assessing them against Lefebvrian notions such as 'the right to difference' and 'the right to the city' (Simonsen 2012, p. 16; see also Simonsen and Koefoed 2020, ch. 4).

As with Longhurst and Johnston, we can see here a desire to produce a nondichotomous understanding of the body and of lived experience. Significantly, this allows the body and emotions/affect to be (re)connected, by collapsing the distinction between inner and outer experience (following Thrift 2008, p. 236). Thus, emotions and affect become different registers through which to understand people's ways of relating to, and interacting with, the world. As much as through bodily senses, emotions and affects are ways that we touch the world and the world touches us (2012, p. 17). Just as, in Merleau-Ponty's famous example, one hand touches the other (1945, p. 106; see Paterson 2007). Thus, Simonsen places the body in the midst of the world, both affecting the world, and also affected by it. Seeing the lived body in its social contexts requires an understanding of the 'sedimented histories' and the 'intersubjective fields' through which bodily life develops. Being in the midst of the world 'renders the body both vehicle and victim of power' (Simonsen 2012, p. 18).

Being in the world, for Simonsen, ultimately implies some kind of common humanity. To be clear, common humanity does not imply harmony, as this is a differentiated world where different bodies have different capacities to have effects upon other bodies (following Ahmed 2000, p. 49). These differential capacities are a product of the techniques through which bodies are differentiated. These involve, for example, the ways that bodies are judged to be familiar or strange or other (see also Ruddick 2010). Thus, Simonsen places the production and experience of otherness at the heart of understanding processes of social differentiation. And, consequently, the politics of critical phenomenology. Being-in-the-world, precisely, immerses the body in struggles over, and for, coexistence. This struggle is the field of the political: where people have different experiences of their differentiated worlds and of their intersubjective fields; where people have differential capacities for affecting change.

In saying this, Simonsen is explicitly thinking about the experiences of migrants, from outside the EU, living in cities in Scandinavia (see also Simonsen and Koefoed 2020, ch. 2). For example, she relates the experiences of Hanif who lives in Copenhagen. Hanif would go out with people from work and school only to find that he was stopped from entering bars and clubs almost all the time. It does not matter whether Hanif is with Danish or coloured people (as he puts it), the experience was repeated. Simonsen reports Hanif as saying 'It was just like getting slapped in the face all the time' (2012, p. 12). Hanif, in Simonsen's understanding, was being othered against the norms of Danish society. Hanif was being 'made strange' through techniques of differentiation that render some bodies familiar and others strange (where familiar/strange is understood primarily through race: see Simonsen and Koefoed 2020, pp. 35–38).

Visible differences are central to the techniques of differentiation: Hanif's skin was read in a way that made him fearful; in Simonsen's words, that made him a 'body suspect' (p. 12). Skin, of course, can be joined with other visible differences, such as clothing, comportment and body shape. And, further, invisible differences, such as voice (accents, loudness) and smell (scents, in foods). For Simonsen, the problem is that Hanif becomes a stranger because the techniques of differentiation only recognise him through racialised grids of meaning and power. This leaves Hanif with feelings of anger and frustration, but with little capacity to effect change. Ultimately, he is left feeling humiliated and excluded. Hanif is not alone in this. Simonsen reports the experience of Abbas taking the train from

Copenhagen to Malmö. In the immediate period after the 2005 bombings of the London Underground, Abbas says:

> you should see their eyes when they see somebody like me enter the train. People really stare [. . .] Then you think, 'shit, I'm the problem here'. If I had just said 'boom', the two persons next to me would have fainted. (Simonsen, 2012 p. 13)

As Simonsen suggests, the resemblance between Abbas' story and Fanon's account of his encounter with a small boy on a train in France is uncanny. She argues that both Abbas and Fanon undergo a visual 'dissection' by the passengers on the train. As we saw in Chapter 2 of this book, in *Black Skin, White Masks*, Fanon argues that black skin becomes the critical site and sight of difference and inferiority in a white-dominated world. Fanon understands that power relations operate through a racialised epidermal schema, which grades bodies according to the colour of the skin. As Fanon moves around the world, he feels that he is 'being dissected by white eyes, the only real eyes' (1952, p. 116). This is not to underestimate the other ways that eyes dissect the body, such as through hair, bone structure, noses, lips, buttocks, gesture and so on; all, chained together to form racialised bodily regimes. And it is not just through the eyes that Fanon is *fixed* into his place within a racialised bodily regime. Seemingly invisible markers, such as blood (and now DNA), sound (voice) and smell, are all added to the toolkit of racialised corporeal schemas.

Nonetheless, it is important to note that the chaining together of different corporeal schemas – of blood and bone, of hair and voice – does not necessarily (or, perhaps, ever) produce bodily regimes that are determinate, consistent and integrated, as Nella Larsen's *Passing* demonstrates (see Chapter 2). Indeed, what race 'is' is worked out in the midst of this confusion. A clear example of this is the fierce debate over Rachel Dolezal's self-description as 'trans racial', which (like other forms of trans identity) calls into question the nature of identity and processes of identification itself. Dolezal refuses the idea that she merely feels herself to be black, but instead insists that she is black. The ontology of race is placed at a remove from processes of identification, whether cultural or biological. Rachel, consequently, occupies the same impossible place as Larsen's Nig. That this impossible place exists is the important point, here. Yet, this impossible place does not delegitimise or prevent the policing of 'race itself', as an epistemic violence. Indeed, rather than trans bodies making bodily regimes inoperable, we can observe that it is precisely because bodies are indeterminate, inconsistent, incoherent and occupy impossible places that they require such force of policing (in Rancière's terms). Force that is, to be sure, both brute and epistemic.

Significantly, Fanon suggests that racialised corporeal schemas put a mirror up to the face of the black man and the reflection tells him that he is inferior and other. The effects of this mirroring are far reaching. Because colonised peoples (can only) recognise themselves as other (to themselves) and because white people are apparently superior and ideal, they are compelled to enact a script that is not their own; to behave according to values and norms that are not theirs; to perform according to standards that they have not set; and, to both identify with and internalise these values, norms and standards as if they were their own. Thus, despite Fanon's constant demand to be (acknowledged as) black, he despairs that 'out of the blackest part of my soul, across the zebra striping of my mind, surges this desire to be suddenly *white*' (p. 63).

By identifying with – and desiring – the position and power of the white man, Fanon argues that the black man ends up by seeing himself in the negative: as not-white, as not-Master and as no-where. Thus, the fractured mirror of corporeal schemas severs Fanon from his own image, from his own body, and spreads him out in front of himself as something not-him. Bodily regimes create bodies by dissection, by overdetermining the meaning of their parts, then putting the body-in-parts back together as a Frankenstein's monster of identity and difference. This corporealisation of the body is not the body, either black or white, as Fanon sees it. As he bitterly observes, 'The Negro is not. Any more than the white man' (p. 231). What is important, then, is where and how the cut between the body and the not-body is drawn.

For Fanon, racialised bodily regime's imposition of skin hierarchies not only defines the visibility of the body, it is also embellished by the white man 'out of a thousand details, anecdotes, stories' (p. 111). These embellishments, which have nothing to do with him, envelope him entirely. Fanon finds himself 'completely dislocated' and 'absolutely depersonalised'. Fanon has been alienated from his self. Fanon is split, both phobia and fetish (see Bhabha 1986, p. 78). And he is aware of being severed from his body. This amputation makes Fanon endure 'a haemorrhage that splattered my whole body with black blood' (p. 112). His eviscerated body is never allowed to be equal to the white man's: across many lines, he suffers the deadly cuts of racist grids of meaning, identity and power. Fanon's body and soul have been incarcerated within the prison-house of white desires, fantasies and fears:

> I discovered my blackness, my ethnic characteristics, – and I was battered down by tom-toms, intellectual deficiency, fetishism, racial defects, slave-ships, and above all, above all else: 'Have a banana'. (p. 112, modified translation)

Fanon is caught between bodily regimes, between the corporeal schemas that seek to define him and to mark him out as different. He finds himself looking at the world and seeing the world look at him. This experience is profoundly dislocating, producing a 'double consciousness' (as Simonsen observes; on this, see also Johnson 2020). This 'double consciousness', perhaps, marks a moment when both inside and outside become visible at the same time, as if suddenly stepping off the Möbius strip of inside/outside and viewing it from afar (which Emmy von N. was never quite able to do, see Chapter 4). In some ways, T. E. Lawrence's experience of being inspected by white eyes also causes him a 'double consciousness'. At one and the same time, he is caught between a colonial bodily regime and an Arab one. Lawrence's experience is also profoundly dislocating, although he is able to laugh at his situation more easily than both Fanon and Abbas. Even so, he is similarly caught between bodily regimes. Lawrence easily adopts and mocks his privileged position as an English officer, yet he also desperately wishes to be recognised as part of the Meccan army.

More importantly, Lawrence wants acknowledgement that he has been recognised as part of the Meccan army by the Sharif of Mecca – and as a Staff Officer, no less. That is, Lawrence wants to see himself reflected in the mirror of the other. Thus, he suffers his dislocation differently than Fanon or Abbas. Lawrence wishes to occupy both locations at once: to be on both sides of the Möbius strip of colonial bodily regimes; to be seen as he wants to be seen, yet to show himself in ways that others cannot understand. A fantasy though this is, this is not a position afforded to either Fanon or Abbas. Instead, they have their bodies handed back to them, dissected and in parts. While Lawrence preserves identity and difference by claiming the place of both 'us' and 'them', 'same' and 'other', Fanon disavows both locations, neither us nor them, same nor other. This is important.

In Möbian logic, the binary locations of same/other and us/them are misleading, whether claimed (Lawrence) or denied (Fanon). Indeed, the idea of 'double consciousness' has itself been criticised for both its binarised logic and its association with Hegelian philosophy (by such writers as Adell 1994), thereby erasing other experiences and also other ontologies. Using Fanon, we might instead ask how double consciousness comes about, how it confines and defines the optics of bodies, and how the friction and struggle contained within the idea of double consciousness might afford opportunities to rethink bodies. Thus, we must seek ways to understand how binaries become operative: how they enable or disable the fixity and flux of similarity and difference, what role they play in the power relations that operate through them. One way that binarised bodily regimes become operative is through emotions and affect: for example, through the desire for, or abjection of, the apparent other. In Abbas' case, Simonsen points out that racialised fears are woven into the fabric of everyday life and into embodied encounters with others (see also Nayak 2011). However, her point is that the emotional and affectual textures of everyday life are intimately connected to power relations that structure people's encounters with others and with the world (see also Crang and Tolia-Kelly 2010). And these power relationships create 'locations', defined by the binary logics of race, class, gender and so on. These locations then become critical sites of contest and struggle in the politics of identity and difference.

The Politics of Location: Power, Space and the Body

Longhurst and Johnston and Simonsen all seek to reassert the importance of the political in thinking geographies of the body. These interventions are also representative of the two dominant approaches to the body within human geography. Broadly, there is an approach that draws on what can be called a 'politics of location' (after Adrienne Rich's signature work, 1984), which further utilises ideas such as positionality and intersectionality to locate people's experiences within the socially produced grids of meaning and power. In addition, there is a phenomenological approach that understands the body as a datum point for experience, often structured through the senses, but especially sight (oftentimes gathered under the heading of visual techniques), touch (primarily deploying the idea of hands that touch one another) and sound (albeit primarily understood through the production of music).

What is clear is that there is currently underway a search for a critical geography of the body that, on the one hand, takes seriously the ways that bodily life is structured by unequal, unjust and injurious social relationships, while at the same time being able to explore people's emotional, affective and lived experiences. In some ways, we can see Longhurst and Johnston and Simonsen as lighting the candle of the same problem, but from different ends. On the one hand, for Longhurst and Johnston, politics emerges out of structured power relationships, where bodies are (in phenomenological terms) differentiated by dichotomies, such as white/black, male/female, masculine/feminine, straight/gay, able-bodied/disabled, sane/mad, rich/poor, Western/non-Western and the like. These power relationships are not necessarily or always apparent, or experienced directly. Consequently, experience itself cannot be the only way to ground or launch politics. On the other hand, for Simonsen, politics emerges from the lived experiences of the differentiated body. Politics is grounded in, and launched by, these the experience of injustice and inequity and addressed by political and ordinary acts of kindness, generosity and hospitality (see Simonsen and Koefoed 2020, ch. 5). The analytical work of understanding the grids of meaning and power that produce these experiences is supplemental to the politics of experience. Taken together, whether politics emerges from structured power relationships or lived experiences, this produces a 'politics of location', where that location is defined by an individual's or group's location within differentiating grids of meaning and power.

However, the 'politics of location' works only because it thinks it knows where people are in the map of meaning and power – and what politics flows from being where they are. In this account, the body becomes a known place, locatable in grids of meaning and power, identity and difference. Effectively, this creates, for everybody, a proper place (as Rancière would call it). This proper place is achieved by a double move. On the one hand, there is an assumption that the grids of meaning and power that create identity and meaning are stable, integrated and coherent and are thereby capable of creating stable, integrated and coherent bodies. On the other hand, the body is converted into a knowable and knowing witnessing point for experience, such that politics can be read off from the body's location within grids of identity and difference. This is what enables Johnston and Longhurst to locate privilege and power in 'white, male, able-bodied, materially well off, Western bodies' (p. 273). This is how Simonsen knows that, in the experiences reported by Hanif and Abbas, it is skin that is the primary marker of difference – and not something else. In both cases, assumptions become conclusions.

However, if we take both Frantz Fanon's and T. E. Lawrence's train journeys together (from Chapters 2 and 3, respectively), the body becomes a less integrated and knowable 'location', not only for Fanon and Lawrence, but also for everyone else. If there are uncanny similarities between Abbas' and Fanon's experiences, there are some parallels between Fanon's and Lawrence's. Both are soldiers, in a foreign country, in a colonial setting. Their identities are mistaken – and read fearfully. Their response shared response, to disguise their anger in humour. Neither are comfortable in their bodies that are being handed back to them by the racial epidermal schemas through which they are being interpolated. Indeed, for both, racial schemas associated with speech and clothing interfere with the epidermal schemas. Both encounters need to be understood as flowing from the antagonism between racial knowabilities and unknowabilities, locations and dislocations.

In Dora and Emmy von N.'s cases, we can similarly see how they struggled with the gendered and sexualised corporeal schemas through which they were meant to live their lives. Indeed, one frame through which they are understood is hysteria as a diagnosis of their symptoms: Dora's coughing (Chapters 5 and 6), for example, and Emmy von N.'s prohibition against being touched whilst also soliciting touch (Chapter 4). What these cases reveal is the topological folding of unconscious material into, through and out of the body-and-mind and the world-as-experienced. I have used the topological form of the Möbius strip to show how the duality of inside and outside is maintained even as they are constantly being over-turned (although the Klein bottle would serve equally well here). This model is also useful for discussing the inside-and-outside nature of unconscious material, which Paul Kingsbury has usefully described as the 'extimacy of space' (2007). However, key to this topological understanding is movement. It is movement across the surface of the Möbius strip (or the Klein bottle) that reveals the flipping of inside and outside. Topological thinking thereby installs the 'trans' (the movement across a seeming divide) at the heart of the production of identity and meaning.

These ideas are do not jettison distinctions between inside and outside (or other dualisms), but rather demand that the distinctions between inside and outside (and other seeming dichotomies) are understood in context. This includes experience (including, when thinking psychoanalytically, the symptom). It also includes structured power relationships. This is to acknowledge that dualistic formulations of identity and difference are the ground for, and product of, social and power relationships (in support of Longhurst and Johnston). However, these distinctions are mobile: at one point, they might seem determinable, yet they constantly flip, get overturned. They simply do not create a single map of identity and difference. And, so, no one way of providing a location for the subject. The politics of the body cannot be read off from the experience of the body itself or its location in the schemas that seek to assign bodies to a proper place.

Indeed, treating bodies as if they can be understood through the schemas that seek to locate them is a 'double oppression' of the body. Thus, the location 'white' only makes sense within a racialised schema for the body that can and does determine who is and is not white. The location 'male' only makes sense if we know what that is and we know who belongs in that category. And so on. Having generated a series of binary schemas, it is then possible to add them up. Creating a map of power and privilege in which everyone's place is knowable and known. And from which politics can be read. Thus, the schemas that ground privilege and power are also deployed in the struggle against those schemas. Yet, using these techniques of identity and differentiation ought also to be brought into question, if only because the identities and differences produced by these techniques are responsible for the very experiences that Simonsen, Longhurst and Johnston wish to struggle against.

The experiences of Fanon, Abbas and Lawrence, of Dora, Emmy von N and other people crazy about their bodies give us pause: if the systems of identity and difference do not 'add up'; if there are locations between or beyond systems of identity and difference, all with contradictory and mutable affects; if the place where people 'locate' themselves – as bodies, as subjects, as thinking-feeling people – is multiple and overdetermined, then how are we to understand the indeterminancies and dislocations from which (I am saying) systems of identity and difference emerge? Let us explore further a couple of ideas that I have deployed in this book: distributions of the sensible and, after that, bodily regimes.

Distributions of the Sensible and the Aesthetic Unconscious

In her paper on Rancière's idea of the distribution of the sensible, Divya Tolia-Kelly explores Rosanne Raymond's engagement with the British Museum's Maori Collections. On one side, she argues, the British Museum should be seen as a political space that reproduces certain civic values and norms. Using Rancière, Tolia-Kelly reveals how the reproduction of these values and norms relies on a particular distribution of the sensible. That is, the British Museum does more than simply provide a framework through which people are expected to think about the cultural artefacts on display, it also organises how its collections are to be sensed and felt about. In doing so, the British Museum produces a distribution of the sensible that correlates a set of abstract conceptual frameworks (such as linear chronological time) with a detached visual regime through which to experience the museum object. Significantly, Tolia-Kelly argues that it is the very specificity of the museum's distribution of the sensible that affords to opportunity to engage with and re-imagine it (similarly, see Pile 2005b). That is, it is possible to draw on aesthetic regimes from elsewhere and bring them into the British Museum not only to highlight the specificity of the museum's distribution of the sensible, but also – through the dissonance between one distribution of the sensible and a distribution of the sensible drawn from elsewhere – to interfere with and challenge the production of senses assumed to be in common.

Critical, in this engagement, is Tolia-Kelly's sense of geo-aesthetics. By thinking aesthetics from elsewhere, as produced in very different places and through different experiences and understandings of the body, Tolia-Kelly is able to see that the distribution of the sensible cannot itself be taken for granted, but rather must be understood through the machines of its production and reproduction, such as the museum. In the British Museum's case, there is a claim to present the world and world history as if it stood outside of the world, presenting objects against a background of abstract regimes of culture, geography and history as if these abstract conceptual regimes were themselves natural and do not themselves serve the social order (to put it in Rancière's terms; see Tolia-Kelly 2019, pp. 127–128). Put bluntly, the British Museum creates a set of universals that organise sensibilities, that frame and fix the cultural object, and naturalises the frameworks through which culture and objects are to be sensed, felt about and understood, using terms such as primitive and modern, beauty and skill, and also the idea of aesthetics itself. Alternatively, for Tolia-Kelly, 'the intervention of postcolonial aesthetics in this scenario disrupts the "naturalised" account of cultural hierarchies and their associated societies' (p. 130).

Tolia-Kelly's discussion of Raymond's challenge to the aesthetic regime of the British Museum does not, therefore, begin with an object in its Maori Collections. Instead, it starts with Raymond's observation that the museum hurts her ears (Tolia-Kelly 2019, p. 131). This is not an idea. This is not just an experience. It is about how Rosanna Raymond is having her body handed back to her as an artefact, as framed within colonial imaginings. What hurts Raymond's ears is the noise of a grass skirt that she feels she has been clothed in. The noise of the grass skirt is unbearable: it is not just that the grass skirt positions her as 'out of place', it also places her 'out of time'. She is dislocated in time and space – and we can easily see the parallels here with Fanon's experience on the train in France. Raymond's body has been cut up and handed back to her, as something pinned to the exotic and the primitive.

Raymond's response is to over-turn the meaning of the grass skirt. She asks whether the swish and swirl of the grass skirt can be made to tell a story of liveliness and vibrancy. For Tolia-Kelly, Raymond's poem *A Throng of Gods* does precisely this. It plays with noise and silence to evoke an alternative postcolonial distribution of the sensible (and here we can see echoes in the Grenfell protests

that use both noise and silence). We are on different ground. Instead of the dominant visual regime, Raymond's and Tolia-Kelly's aesthetics require a different
mode of hearing and of touch. And this itself interferes with the visual: instead of
the visual as a detached regime that is intended to prompt thought and reflection,
it is enwrapped in an embodied regime intended to prompt noise and feeling.
This forms a new fabric of the aesthetic, for Tolia-Kelly:

> Raymond's greatest 'weapon' in Goddess form is her powerful use of the senses and
> sensibilities that are at odds with each other. (2019, p. 133)

For Tolia-Kelly, Raymond's new aesthetic aligns with Ranière's demand that
aesthetics should 'articulate affective politics and demonstrate new ways of
"doing" progressive politics' (2019, p. 123; following Rancière 2010; see also
Tolia-Kelly 2016). The postcolonial intervention in dominant aesthetic regime
creates opportunities not only for re-thinking the production of common sense
within the British Museum, the dissonance between Raymond's aesthetic regime
and the British Museum's also makes clear how these aesthetic regimes – how
their distributions of the sensible – are at odds within one another.

For Tolia-Kelly, this at-odds-ness creates the possibility of an unbounded subject, untethered from a time and a place. This free subject is, further, disruptive
of the means of fixing subjects in place and time. I have also sought to develop
this idea of a subject marked by senses and sensibilities that are at odds with one
another. However, following Freud, my unbounded subject is not free; nor can
the coexistence of senses and sensibilities be used as a form of counter-essen-
tialism that places the subject outside of determinations of all kinds. Instead, the
at-odds-ness of the subject is a product of both the indeterminacies and overdeterminations of meaning and power. For this reason, I have had to alter Rancière's
notions of the distribution of the sensible and of aesthetics. There are two moves
to achieve this. First, in line with Tolia-Kelly, it is important to understand
distributions of the sensible in the plural and to see them as malleable. Tolia-Kelly
shows that distributions of the sensible exist elsewhere; these elsewheres, as significantly, are also here (in this case, in London). Second, the distribution of the
sensible cannot be thought of as a coherent and integrated regime of aesthetics.
The aesthetic regime is itself an attempt to render contradictory senses, feelings
and ideas into some kind of order, with policings of different kinds, whether
through museums or through housing policy or through the police (see next
section).

As we saw in Chapter 1, for Rancière, the aesthetic is an unconscious regime
that organises the sensible. In my understanding, aesthetic forms bear witness to
unconscious mechanisms designed to give expression to, and to manage, unconscious dynamics. Rather than a unitary and lifeless understanding of the unconscious, I have built a model of the unconscious based in its repressive, distributive
and communicative functions (in Chapters 5 and 6) by extending and reworking

Freud's account of the unconscious. In doing so, we can see that the unconscious is not only worldly, but is differentially engaged with the world through its different functions, even contradictorily and ambivalently so. Thus, we have seen that the repressive functions of the unconscious easily align with what Rancière would call the social order and with the aesthetic regime, enabling the internalisation of cultural values and norms as if they were one's own (see, e.g., Chapter 5). However, we have also seen that repression runs counter to the imposition of the social order, producing traumatic experiences and performativities of the body that discompose social norms and values (see, e.g., Chapter 4).

In this understanding, aesthetic regimes are characterised by attempts to impose meanings many times over in the face of structural indeterminacy. They are challenged not just from elsewhere, but by elsewheres within: elsewheres within the psyche and the social. Or, put another way, the idea of the aesthetic unconscious underscores that the subject is radically open, already constituted by its engagements with the world. These engagements do not render the subject coherent and integrated, but rather indeterminate and overdetermined. The struggle, then, is to achieve coherence and integration – and it is this struggle that prompts social orderings and policings. A struggle that prompts institutions, such as the British Museum, to enable people to know where they are in the great scheme of things. This scheme of things does not have to be singular nor unchanging to reflect the social order, but it does have to be policed. Indeed, I have argued, it is because indeterminacy and overdetermination are constitutive of the social order that the social order requires policing. Or, put another way, it is because bodily regimes do not enfold bodies that bodies must be policed, paradoxically by operationalising bodily regimes.

Bodily Regimes and Affective Politics

In the last section, and in Chapter 7, we saw how artistic interventions can seek to disrupt dominant distributions of the sensible that are underpinned by an aesthetic unconscious (which is no longer thought of as static or singular). Indeed, for Kivland (in Chapter 7), we saw that her interventions are actually designed to utilise and operate on unconscious contents and by activating unconscious processes. For Tolia-Kelly, Rowland's disruptions come from outside the dominant aesthetic regime, while for Kivland disruption in the aesthetic regime is generated by utilising the inherent contradictions inside the social order. The importance of (both) these kinds of disruptions is that they are, for Rancière, politics as such. For Rancière, Mustafa Dikeç argues, politics 'implies a relationship between aesthetic forms – as objects of sensory perception – and established orders of hierarchy and domination' (2015, p. 10). Significantly, politics is what causes a disruption in the aesthetic regime that creates and maintains the social order.

Following Rancière, Dikeç argues that politics is incendiary, designed to raze the aesthetic regime that supports the social order. Politics 'defeats our senses and disrupts our ways of sensing and making sense of the world' (2015, p. 11). Thus, politics 'unsettles our routinised ways of sensing and making sense of the world by introducing an element that cannot initially find a register in our habitual spatial orderings' (p. 11). Like Tolia-Kelly, Dikeç searches for this disruptive element in spaces outside the social order. Instead of alighting upon an aesthetic form, however, Dikeç focus upon an affect: rage. More precisely, he focuses on the spatialisation and politicisation of an affect: urban rage.

In *Urban Rage*, Dikeç shows how urban policies, policing of various kinds and the police are used to exclude and marginalise racialised subjects spatially by assigning them to, what in Britain are commonly described as, 'left behind' places and spaces (2017, which broadens his analysis of Paris' *banlieues*, 2007). Places can be neighbourhoods; spaces can be tower blocks. The social order polices bodies brutally, ensuring that they stay in their proper place. This policing is not only achieved by the police, but the police have nonetheless become increasingly militarised as have urban spaces themselves (see Graham 2010). From this perspective, the urban social order is reproduced through bodily regimes that mark bodies to ensure that they can be and are policed, whether through housing policy, surveillance or profiling, whether through skin or clothing or behaviours.

There is an affective politics to this. The brutality associated with policing bodily regimes, usually by the police, Dikeç shows, creates resentment and anger. Over time, these affects build – and explode in urban uprisings. This process, for Dikeç, is ordinary. It keeps happening. In different cities. In different historical periods. Urban uprisings are to be expected, because the policing of the social order through bodily regimes creates an outside from which people can launch politics as such – an incendiary moment that threatens to wreck the social order. An example is the London Riots of 2011 (see Dikeç 2017, ch. 3).

On 4 August 2011, Mark Duggan was killed by armed police in Tottenham, north London. The history of Tottenham – like that of Notting Dale (see Chapter 1) – is filled with tension between black Londoners and the Metropolitan Police (hereafter, the Met). For example, there had been riots in Tottenham in 1985 on the Broadwater Farm Estate, less than a mile and a half from where Mark Duggan was shot. Along with this history, Mark Duggan's death was seen as yet another murder of black people at the hands of the police. Indeed, the 1985 Broadwater Farm Estate riots were prompted by the deaths of two black women during police searches: Cynthia Jarrett and Cherry Groce. In the years leading up to the Mark Duggan incident, the Met had been involved in a series of controversial deaths of black and minority ethnic people: such as (but not only) Joy Gardner in 1993, Oluwashijibomi 'Shiji' Lapite in 1994, Brian Douglas in 1995, Ibrahima Sey in 1996, Sarah Thomas in 1999, Jean Charles de Menezes in 2005, and Sean Rigg in 2008. In each case, it seemed the authorities would not, or were reluctant to, find that the Met had done anything wrong.

In the days after Mark Duggan's death, mistrust between the black community and the Met was fuelled by rumour and counter-rumour: there was a story about a bullet being embedded in a police radio, implying that Duggan had shot at police, while Duggan's family insisted he was not armed. On Saturday 6 August a peaceful protest march organised by Duggan's family, starting at the Broadwater Farm Estate, completed its journey at Tottenham Police Station, less than a mile away. A crowd of about 300 gathered outside, demanding justice for Duggan. 'We want answers', they chanted. Then, a rumour appeared on social media that a 16-year-old girl had sustained injuries while attacking the police with a champagne bottle. Whether this was the trigger or not (or even true or not), by the late evening riots and looting were taking place in Tottenham. That night, the looting moved from Tottenham to Wood Green High Road (on the other side of the Broadwater Farm Estate) and down towards Turnpike Lane tube station, with the police unable to stop it.

On Sunday 7 August there were outbreaks of rioting and looting across London. By Monday 8 August, not just London was burning; almost every city in England seemed to be in flames. It was not until Wednesday 10 August that police forces across the country seemed to gain any kind of control. Indeed, urban uprisings in 2011 were not confined to England. An incendiary politics seemed to be setting the world alight. There was also the Arab Spring and the Egyptian Revolution. By September, there would be the Occupy movement, starting in Zuccotti Park in New York and spreading quickly to other cities around the world, from Canada to Colombia, from Taiwan to Turkey (for a Rancièrian analysis, see Bassett 2014). Rage seemed everywhere. The affective politics of rage burning down the social order.

For Dikeç, 'the rage that erupts into uprisings is the flip side of [an] urban age' characterised by unprecedented 'levels of inequality' (2017, p. 2). In this light, the affective politics of rage is a product of the social and spatial order: the inevitable outcome of hierarchies, domination and brutality. Indeed, cities themselves become agents of revolution (Harvey 2013), creating the possibilities for dispute, disruption and dissensus. Thus Dikeç argues, riots are:

> spontaneous insurrections, not planned events or organised movements, which are motivated by grievances that have to do with everyday urban lives of those who revolt: those who live on the wrong side of town fearing a brush with the police might put an end to their lives; those who are on the right side of town but have the wrong colour; those who find the wrong side of town where they live is now becoming the right side, squeezing them out; those, in general, whose everyday urban lives are marked by reminders of their exclusion from the wealth, rights and privileges available to other urban dwellers. Urban rage that erupts in uprisings is the revolt of the excluded, overtaking urban spaces through unruly practices and defying the order of things. (2017, p. 3)

It is hard to disagree. Yet, what we learned from *Handsworth Songs*, a 1986 film about the 1985 riots in Liverpool and London by the Black Audio Film Collective, is that 'there are no stories in the riots, only the ghosts of other stories'. Dikeç converts the story of the 2011 riots into rage and rage into the affective politics of the excluded. This is only one story. In fact, it is always the story of riots. If there is a Rancièrian element in the riots that disrupts bodily regimes, then is it to be found in 'rage' – the all too familiar and commonplace affect of radical politics? (As an aside, it is also worth noting that rage is not the preserve of radical or progressive politics.) There is some work that suggests that radical politics does not have to rely on rage, but on other kinds of political affinity building, for example, on the use of parody (see Routledge 2012) and laughter (see Noxolo 2018), or on forms of solidarity (see, e.g., Featherstone 2007). While rage can be busy reinforcing the division between acceptable and unacceptable behaviour, between those who can speak (the authorities) and those who cannot (rioters), acts of care, community and solidarity can achieve the exact opposite: a quieter (or differently noisy), less spectacular, re-drawing of the division between those who are seen and heard and those are not.

In the wake of the first night of the Tottenham riots, Wood Green High Street had been taped off by the police. The entire street seemed to be a scene of crime. Yet, people wandered freely up and down the road, looking at the night's destruction. A half burned-out Vauxhall Astra. A smoke blackened shop. Broken windows. Sheets of shattered safety glass. The remains of looted goods, fallen or torn from hands, trod underfoot: not worth the effort of recovery. On the street, an elderly black woman was telling people her experience. I asked her what happened. She had been heading home on the 123 bus that rioters had set ablaze. She explained how terrifying the experience was. Then she moved on to tell someone else. A stranger nodded to me: 'terrible, just terrible', he said. Others agreed. As much as the riots might be the authentic expression of the anger and frustration of marginalised youth, there were others in the community for whom this was an affront. A boarded-up window in The Body Shop became an impromptu message board: the messages, written on colourful Post-it Notes®, were mainly of love and community (see Figure 8.1).

'Let's all live and love together', one read. Another said, 'we stand together as a community'. Someone wrote: 'We all need to smell good' – perhaps a reference to the widespread looting of high-end perfumes (and other beauty products) that night. Such sentiments are not counter-revolutionary. Perhaps the real disruptions in the uprisings were in the other claims to speak for place and space, ones that rely on love and community. Again, to be sure, this is not to romanticise the idea of community (there's no single story in the riots), but instead to suggest that rage is not the only structuring principle for affective politics and not the only response to the clash of bodily regimes.

Rage, in itself, can be an entirely safe place for politics: assigning angry bodies to a particular space and place, where they can be policed and marginalised, all over again. As Dikeç says, urban rage keeps happening: it perpetually returns, seemingly from the outside, only to disappear – as if following a never-ending line

Figure 8.1   The Body Shop's wall of love, August 2011. Source: Steve Pile.

along a Mobius strip, occasionally flipping from outside to inside and then back again. Where is the cut in the political Mobius strip that transforms the changing same into history? This question does not undermine the importance of political struggles that emphasise the affective politics of rage or the political possibilities generated by urban uprisings. Rather, this is to pose a question about whether there's more to the politics of the body than locating people's position (whether structural or experiential) within geometric grids of meaning and power, creating straight lines, for example, between excluded bodies, rageful affective politics and the overturning of the social order. Rather, I have sought an alternative to this power geometry. This alternative emphasises the ways that bodies are produced as bodies through imbricated bodily regimes – bodily regimes that never completely enfold the body or people's experiences. Indeed, people live everyday with the experience of being between bodily regimes. Thus, it is also the common place experience of the clash of bodily regimes that produces politics. This brings us back to Grenfell – and the injuries of bodily regimes and of politics.

The Clash of Bodily Regimes: Struggling with Location, Location, Location

One of the largest post-war redevelopment schemes in the (newly formed) London Borough of Camden, the Chalcots estate was built over a five-year period between 1965 and 1970; two years before work began on the Lancaster West Estate in Kensington. Unlike the Notting Dale development, it did not require slum clearance, as German bombs (intended for nearby railway lines) had already demolished much of the existing housing. Already flattened, the site seemed perfect for a grand plan. Albeit on not such a grand scale, Camden's chief architect Sydney Cook drew heavily on Le Corbusier's 1925 Plan Voisin. The plan envisaged a council estate with five high-rise tower blocks connected by low-rise finger blocks. The design included residential gardens and facilities for young people. Uncompromisingly modern in its design, it was intended to be uncompromisingly democratic in its design for living. Four identical high-rise buildings of 23 storeys were built along with a fifth, shorter, 19-storey block, providing a combined total of over 700 flats. Between 2007 and 2019, the original concrete facades were clad with metal panels. The panels were similar to those used on the Grenfell Tower. In the wake of the fire, Camden Council quickly acted to remove the cladding, evacuating residents in the process (albeit without consultation or consent). When I walked around the estate in July 2019, it was impossible to miss the signs: '#Justice4Grenfell' and '#neveragain', they read (see Figure 8.2).

On 23 November 2018, the *Independent* newspaper reported research that had discovered that 96% of London's high-rise council-owned tower blocks are without a sprinkler system. Since the Grenfell fire, the London Fire Chief has made repeated calls (e.g. on 13 September 2017, 8 July 2018 and 11 February 2019) for there to be intervention on, and legislation to reinforce, the use of sprinkler systems in high rise buildings. Councils themselves have costed the retro-fitting of sprinkler systems in their housing stock at hundreds of millions of pounds and demanded the cost be met by central government, as promised. On 9 May 2019, the government announced it would give Councils £200 million to replace Grenfell-style cladding. The money averages out at about £1.2 million per affected building, with the true costs of replacement estimated at between £4 and £5 million.

Grenfell United, a campaign group for survivors and the bereaved, welcomed the move. It was, they said, a small step. They added: 'the truth is we should never have had to fight for it' (quoted in 'Government Allocates £200 million to Replace Grenfell-style Cladding', *The Guardian*, 9 May 2019). After a year-long campaign, bowing to public pressure, the government eventually made £1 billion available to replace aluminium composite cladding on all private and social housing over 18 metres high. Yet, by 6 March 2020, according to the UK Parliament, out of

Figure 8.2   '#Justice4Grenfell' on railings outside Dorney Tower Block on the Chalcots Estate, Camden, January 2019. Source: Steve Pile.

175 private sector properties, no remedial work had started on 143 of them, while remediation work on only 45% of social housing properties had been completed (69 out of 155), with the National Housing Federation estimating that the eventual cost of removal for the social housing sector alone will be about £10 billion. In some ways, we might see this as a kind of selective deafness on part of government in response to criticisms and activism. Dan Renwick sums up the situation this way:

> Across the country, people are organising to be heard. They are seldom recognised for their actions, but when they are, their moving image is taken, but their voices have been kept on mute, apart from a few selected sound-bites according with a pre-established script. (2019, p. 43)

Cladding is not the only way to measure the urgency and commitment of the government's response. To the dismay of the survivors and relatives, on 7 March 2019 the Metropolitan Police announced that there would be no prosecutions associated with the fire until after the conclusion of the public inquiry; so, not until 2022 at the earliest, despite the police saying in June 2019 that they had identified suspects for possible manslaughter charges. This echoes the decision in August 2018 not to prosecute the most senior police officer involved in the Hillsborough disaster, Chief Constable Norman Bettison: after almost 30 years, witness testimonies, the Crown Prosecution Service admitted, have become unreliable.

This book began with the opening of the inquiry into the Grenfell Tower fire. The first days were full of fire, the fury at a social injustice measured in people's lives, dying horrifically in an avoidable tragedy. Bodies. Affects. Politics. Entangled in a demand for truth and justice. Since then, the politics of Grenfell has slowly slipped into a discussion about socio-technical systems of fire prevention, their effectiveness and their cost. About cladding, and its combustibility. About who installed it, and what they knew (and did not know) about the fire risks of different cladding types. About regulatory systems, and their enforcement. This disentangles affects from decisions, whose costs are now measured in pounds not lives. The many stories of the fire are slowly converted into a debate about the response of the fire brigade on the day and upon fire prevention in high-rise buildings. The politics of Grenfell is becoming focused upon the fire brigade and upon Council spending limits and priorities. Justice, itself, focused upon prosecutions, upon legal procedures. Where the rallying cry for Justice once called into question what Justice looks like, it is slowly being converted back into politics-as-usual.

Thus, the increasing focus on socio-technical systems shifts the terrain upon which justice is understood. Justice becomes the attribution of blame and the application of punishment within a legal framework. The same framework that failed the Hillsborough families. By emphasising socio-technical questions, ideas of justice associated with social inequality and economic marginalisation become increasingly marginal. And the debate about what justice looks like disappears, while the affective politics of rage and loss, of compassion and solidarity, is slowly de-intensified. This changes politics. Arguably, this narrows the field of politics onto questions of cladding removal and regulation, onto changes in fire brigade operating procedures and equipment, and onto the imposition of jail sentences on the guilty.

We can now observe that what has also happened, through the focus on socio-technical systems, is a shift in whose voices are heard, whose bodies are seen and what affects are deemed significant. Thus, the increasing significance of technical and legal questions becomes a measure of the reassertion of a distribution of the senses that emphasises reason over affect, experts over experiences, equipment and regulation over practices. And, in this shift, a certain de-politicisation and de-affectualisation is performed. This is not apolitical or post-political, nor is it unaffecting or unemotional. Rather, this process slowly removes the Grenfell tragedy from the domain of affective politics through which contemporary politics becomes operative. An affective politics that, ironically, asserts affect over reason, experiences over experts, fantastic technical solutions over real world dilemmas: a political Möbius strip, which can flip and fudge logics at will, yet seem consistent – and remain powerful. As we have seen, the aesthetic unconscious of affective politics is operative *because* it can contain opposites, can determine meaning of things many times over and work with the indeterminacy of everyday life. It polices bodies, through bodily regimes that assign bodies to particular places, such as Grenfell Tower and the Broadwater Farm Estate, not because they are known or knowable, but because they are not.

Grenfell created a moment when the fractures between bodily regimes enabled more voices to be heard, more bodies to be seen and taken account of, and the expression of anger and pain to be transmitted and felt far beyond those immediately impacted by the fire. People located, located, located in bodily regimes associated with class, race, gender, sexuality and age. . .with migration, housing and work. . .were, in the aftermath of the fire, suddenly at odds with these frames of reference. Instead, new communities of action, affect and politics formed around an outpouring of generosity, welcoming and mutual aid. What brought people together was not an already existing shared identity, nor an already existing fully operative ideal of community. Rather, these already existed in parts. Politics emerged not from the singularity of community, but from its crystallisation out of alliances formed along different lines of solidarity and action. Just as there are many stories in the fire of Grenfell, so there are many emergent forms of politics. This is important. Some form around noise (as in traditional protest marches), others around silence (as in the monthly Grenfell silent walks and remembrance events). These are not mutually exclusive: they fold in and out of one another. They are not inevitable or permanent either.

The 'politics of location', in some ways, relies upon the determinacy and permanence of identity and difference. For me, any consistency of identity is an achievement – a struggle to achieve. Identity is not a permanent location. Nor is it determinate. Nor localisable. Systems of differentiation that produce grids of meaning and power seek to locate and fix bodies within them. Yet, throughout this book, we have seen consistently that people live with the clash of bodily regimes. This is not comfortable. Too often, it can be traumatic and injurious. But the spaces between bodily regimes are nonetheless a location for a politics of the body. A location that we now know does not add up or settle down: dis-locations that are mutable and indeterminate. Indeed, it is the multiplicity, indeterminacy and mutability of these *between bodily regime* locations that demonstrates that the body can be lived differently.

References

Adell, S. (1994) *Double-Consciousness/Double Bind: Theoretical Issues in Twentieth Century Black Literature*. Urbana: University of Illinois Press.

Ahmed, S. (1999) '"She'll wake up one of these days and find she's turned into a nigger": Passing through hybridity', *Theory, Culture and Society.* 16(2), pp. 87–106.

Ahmed, S. (2000) *Strange Encounters: Embodied Others in Post-coloniality*. London: Routledge.

Ahmed, S. and Stacey J. (eds.) (2001) *Thinking through the Skin*. London: Routledge.

Allen, J. (2003) *Lost Geographies of Power*. Oxford: Basil Blackwell.

Allen, J. (2011) 'Topological twists: Power's shifting geographies', *Dialogues in Human Geography.* 3(1), pp. 283–298.

Allen, J. (2016) *Topologies of Power: Beyond Territory and Network*. London: Routledge.

Anderson, B. (2009) 'Affective atmospheres' *Emotion, Space and Society*, 2(2), pp. 77–81.

Anzieu, D. (1985) *The Skin Ego*. London: Karnac Books, 2016.

Appiganesi, L. and Forrester, J. (2005) *Freud's Women*. Second Edition, London: Phoenix.

Apter, E. (1991) *Feminizing the Fetish: Psychoanalysis and Narrative Obsession in Turn-of-the-century France*. Ithaca: Cornell University Press.

Baker, H. A. (1987) *Modernism and the Harlem Renaissance*. Chicago: Chicago University Press.

Barr, J. (2006) *Settling the Desert on Fire: T. E. Lawrence and Britain's Secret War in Arabia, 1916–1918*. London: W. W. Norton and Company, 2008.

Bassett, K. (2014) 'Rancière, politics, and the Occupy movement', *Environment and Planning D: Society and Space.* 32, pp. 886–901.

Bell, D. and Valentine, G. (eds.) (1995) *Mapping Desire: geographies of sexuality*. London: Routledge.

Bennett, J. (1996) *The Passing Figure: Racial Confusion in Modern American Literature*. New York: Peter Lang.

Bodies, Affects, Politics: The Clash of Bodily Regimes, First Edition. Steve Pile.
© 2021 Royal Geographical Society (with the Institute of British Geographers).
Published 2021 by John Wiley & Sons Ltd.

Benthien, C. (2002) *Skin: On the Cultural Border between Self and the World*. New York: Columbia University Press.

Berger, J. (1972) *Ways of Seeing*. Harmondsworth: Penguin Books.

Bernheimer, C. and Kahane, C. (eds.) (1990) *In Dora's Case: Freud-Hysteria-Feminism*. Second Edition. New York: Columbia University Press.

Bhabha, H. (1994) *The Location of Culture*. London: Routledge.

Bick, E. (1968) 'The experience of the skin in early object relations'. In: *Surviving Space: Papers on Infant Observation* (ed. A. Briggs), 55–55. London: Karnac Books, 2002.

Bingley, A. (2003) 'In here and out there: sensations between Self and landscape', *Social & Cultural Geography*. 4, pp. 329–345.

Blackman, L. (2007) 'Reinventing psychological matters: the importance of the suggestive realm of Tarde's ontology', *Economy and Society*. 36(4), pp. 574–596.

Blackman, L. (2008) 'Affect, relationality and the problem of personality', *Theory, Culture and Society*. 25(1), pp. 27–51.

Blackman, L. (2010). 'Embodying affect: voice-hearing, telepathy, suggestion and modelling the non-conscious', *Body and Society*. 16(1), pp. 163–192.

Blanton, D. and Cook, L. M. (2002) *They Fought Like Demons: Women Soldiers in the American Civil War*. Baton Rouge: Louisiana State University Press.

Blum, V. and Secor, A. (2011) 'Psychotopologies: Closing the circuit between psychic and material spaces', *Environment and Planning D: Society and Space*. 29(6), pp. 1030–1047.

Blum, V. and Secor, A. (2014) 'Mapping trauma: Topography to topology'. In: *Psychoanalytic Geographies* (eds. P. Kingsbury and S. Pile), 103–118. London: Routledge, 2016.

Bohm, D. (1990) 'A new theory of the relationship of mind and matter', *Philosophical Psychology*. 3(2), pp. 271–286.

Bonnett, A. (2000) *White Identities: Historical and International Perspectives*. London: Prentice Hall.

Borch, C. (2006) 'Urban imitations: Tarde's sociology revisited', *Theory, Culture and Society*. 22(3), pp. 81–100.

Boyce-Davies, C. (2008) *Left of Karl Marx: The Political Life of Black Communist Claudia Jones*. Durham, N.C.: Duke University Press.

Bradley, G. M. (2019) 'From Grenfell to Windrush'. In: *After Grenfell: Violence, Resistance and Response* (eds. D. Bulley, J. Edkins and N. El-Enany), 135–142. London: Pluto Press.

Brennan, T. (2004) *The Transmission of Affect*. Ithaca: Cornell University Press.

Bressey, C. (2009) 'The legacies of 2007: remapping the black presence in Britain', *Geography Compass*. 3(3), pp. 903–917.

Bressey, C. (2014) 'Archival interventions: Participatory research and public historical geographies', *Journal of Historical Geography*. 46, pp. 102–104.

Breuer, J. (1895) 'Fräulein Anna O.' In: *Studies in Hysteria* (S. Freud and J. Breuer), 25–50. Harmondsworth: Penguin Modern Classics, 2004.

Brody, J. D. (1992) 'Clare Kendry's "true" colors: Race and class conflict in Nella Larsen's *Passing*', *Callalo*. 15(4), pp. 1053–1065.

Bromberg, P. M. (2009) 'Hysteria, dissociation and cure: Emmy von N revisited', *Psychoanalytic Dialogues*. 6(1), pp. 55–71.

Brottman, M. (2011) *Phantoms of the Clinic: From Thought Transference to Projective Identification*. London: Karnac Books.

Browne, K., Nash, C. and Hines, S. (2010) 'Introduction: Towards trans geographies', *Gender, Place and Culture*. 17, pp. 573–577.

Brubaker, R. (2016a) 'The Dolezal affair: Race, gender, and the micropolitics of identity', *Ethnic and Racial Studies*. 39(3), pp. 414–448.

Brubaker, R. (2016b) *Trans: Gender and Race in an Age of Unsettled Identities*. Princeton: Princeton University Press.

Butler, J. (1993) *Bodies that Matter*. London: Routledge.

Butler, R. and Parr, H. (eds.) (1999) *Mind and Body Spaces: Geographies of Illness, Impairment and Disability*. London: Routledge.

Callard, F. (2003) 'The taming of psychoanalysis in geography', *Social & Cultural Geography*. 4, pp. 295–312.

Campbell, J. (2006) *Psychoanalysis and the Time of Life: Durations of the Unconscious Self*. London: Routledge.

Campbell, J. (2007) 'Transference streams of affects and representations', *International Journal of Critical Psychology*. 21, pp. 50–75.

Campbell, J. and Pile, S. (2010) 'Telepathy and its vicissitudes: Freud, thought transference and the hidden lives of the (repressed and non-repressed) unconscious', *Subjectivity*. 3(4), pp. 403–425.

Campbell, J. and Pile, S. (2011) 'Space travels of the Wolfman: Phobia and its worlds', *Psychoanalysis and History*. 13(1), pp. 69–89.

Campbell, J. and Pile, S. (2015) 'Passionate forms and the problem of subjectivity: Freud, Frau Emmy von N. and the unconscious communication of affect', *Subjectivity*. 8(1), pp. 1–24.

Carby, H. (1987) *Reconstructing Womanhood: The Emergence of the African American Woman Novelist*. Oxford: Oxford University Press.

Carter, P. (2006) 'The penumbral spaces of Nella Larsen's *Passing*: Undecidable bodies, mobile identities, and the deconstruction of racial boundaries', *Gender, Place and Culture*. 13(3), pp. 227–246.

Cavanagh, S., Failler A. and Hurst, R. (eds.) (2013) *Skin, Culture and Psychoanalysis*. Basingstoke: Palgrave Macmillan.

Charles, M. (2019) 'ComeUnity and community in the face of impunity'. In: *After Grenfell: Violence, Resistance and Response* (eds. D. Bulley, J. Edkins and N. El-Enany), 167–192. London: Pluto Press.

Cheng, A. (2010) *Second Skin: Josephine Baker and the Modern Surface*. Oxford: Oxford University Press.

Cixous, H. (1976) *Portrait of Dora*. London: Routledge, 2004.

Clough, P. T. and Halley, J. (eds.) (2007) *The Affective Turn: Theorizing the Social*. Durham, N.C.: Duke University Press.

Cohen, R. (2016) 'The spatiality of being: Topology as ontology in Lacan's thinking of the body'. In: *Psychoanalysis: Topological Perspectives* (eds. M. Friedman and S. Tomšic, 227–250. Bielefeld: transcript Verlag.

Conan Doyle, A. (1926) *The History of Spiritualism (complete)*. London: The Echo Library, 2006.

Connor, S. (2003) *The Book of Skin*, London: Reaktion.

Crang, M. and Tolia-Kelly, D. (2010) 'Guest editorial: Affect, race, and identities. Visceral, viscous theories of race after social constructivism', *Environment and Planning A*. 42, pp. 2309–2314.

Crang, P. (2010) 'A curious mapping of American culture', *Social & Cultural Geography.* 11(8), pp. 922–923.

Creighton, M. and Norling, L. (eds.) (1996) *Iron Men, Wooden Women: Gender and Seafaring, 1700–1920.* Baltimore: John Hopkins University Press.

Crookes, W. (1874) 'Notes of an enquiry into the phenomena called spiritual, during the years 1870–1873', *Quarterly Journal of Science,* January. Reprinted in W. Crookes, *Researches in the Phenomena of Spiritualism,* 81–102. London: J Burns, 1874.

Danewid, I. (2019) 'The fire this time: Grenfell, racial capitalism and the urbanisation of Empire', *European Journal of International Relations,* early view online, 25 pp.

Davis, T. M. (1994) *Nella Larsen, Novelist of the Harlem Renaissance: A Woman's Life Unveiled.* Baton Rouge: Louisiana State University Press.

Davis, W. (1995) *Dreaming the Dream of the Wolves: Homosexuality, Interpretation and Freud's 'Wolfman'.* Indianapolis: Indiana University Press.

Dawahare, A. (2006) 'The gold standard of racial identity in Nella Larsen's *Quicksand* and *Passing*', *Twentieth-Century Literature.* 52(1), pp. 22–41.

Dawson, A. (2007) *Mongrel Nation: Diasporic Culture and the Making of Postcolonial Britain.* Ann Arbor: University of Michigan Press.

Dawson, G. (1994) *Soldier Heroes: British Adventure, Empire and the Imagining of Masculinities.* London: Routledge.

De Certeau, M. (1980) 'Utopies vocales: glossolalies', *Traverses.* 20, pp. 16–37.

de Noronha, N. (2019) 'Housing policy in the shadow of Grenfell'. In: *After Grenfell: Violence, Resistance and Response,* 2019 (eds. D. Bulley, J. Edkins and N. El-Enany), 143–164. London: Pluto Press.

De Marneffe, D. (1991) 'Looking and listening: The construction of clinical knowledge in Charcot and Freud', *Signs.* 17(1), pp. 71–111.

Decker, H. S. (1991), *Freud, Dora and Vienna 1900.* New York: The Free Press.

Delaney, D. (2002) 'The space that race makes', *The Professional Geographer.* 54(1), pp. 6–14.

Deleuze, G. (1998) *Essays Critical and Clinical.* London: Verso.

Diamond, N. (2013) *Between Skins: The Body in Psychoanalysis – Contemporary Developments.* London: John Wiley and Sons.

Didi-Huberman, G. (1982) *Invention of Hysteria: Charcot and the Photographic Iconography of the Salpêtrière.* Cambridge, Mass.: The MIT Press, 2003.

Dikeç, M. (2005) 'Space, politics, and the political', *Environment and Planning: Society and Space.* 23, pp. 171–188.

Dikeç, M. (2007) *Badlands of the Republic: Space, Politics and Urban Policy.* London: Blackwell.

Dikeç, M. (2015) *Space, Politics and Aesthetics.* Edinburgh: Edinburgh University Press.

Dikeç, M. (2017) *Urban Rage.* New Haven: Yale University Press.

Dimen, M. and Harris, A. (eds.) (2001) *Storms in Her Head: Freud and the Construction of Hysteria.* New York: Other Press.

Doherty, G. (2018) *Grenfell Hope: Ravaged by Fire but Not Destroyed.* London: SPCK Publishing.

Duncan, N. (ed.) (1996) *Bodyspace: Destabilising Geographies of Gender and Sexuality.* London: Routledge.

Dwyer, O. J. and Jones, J. P. (2000) 'White socio-spatial epistemology', *Social & Cultural Geography.* 1(2), pp. 209–222.

Dyer, R. (1997) *White: Essays on Race and Culture*. London: Routledge.

El-Enany, N. (2019) 'Before Grenfell: British immigration law and the production of colonial spaces'. In: *After Grenfell: Violence, Resistance and Response* (eds. D. Bulley, J. Edkins and N. El-Enany), 50–61. London: Pluto Press.

Ellenberger, H. (1970) *The Discovery of the Unconscious: The History and Evolution of Dynamic Psychology*. New York: Basic Books.

Ellenberger, H. (1977) 'The story of "Emmy von N."': A critical study with new documents'. In: *Beyond the Unconscious: Essays of Henri F. Ellenberger in the History of Psychiatry* (eds. M. S. Micale), 273–290. Princeton: Princeton University Press, 1993.

Emerson, S. (1988) *Secret Warriors: Inside the Covert Military Operations of the Reagan Era*. New York: Penguin.

Eshel, O. (2006) 'Where are you, my beloved? On absence, loss, and the enigma of telepathic dreams', *International Journal of Psychoanalysis*. 87, pp. 1603–1627.

Esson, J., Noxolo, P., Baxter, R., Daley, P. and Byron, M. (2017) 'The 2017 RGS-IBG chair's theme: Decolonising geographical knowledges, or reproducing coloniality?' *Area*. 49(3), pp. 384–388.

Everett, A. (2018) *After the Fire: Finding Words for Grenfell*. Norwich: Canterbury Press.

Fanon, F. (1952) *Black Skin, White Masks*. London: Pluto Press, 1986.

Featherstone, D. (2007) *Resistance, Space and Political Identities: The Making of Global Counter-global Networks*. Chichester and Oxford: Wiley-Blackwell.

Flammarion, C. (1892) *La Planète Mars et ses Conditions d'Habitabilité*. Paris: Gauthier-Villars.

Flournoy, T. (1899) *From India to the Planet Mars: A Case of Multiple Personality with Imaginary Languages*. Princeton: Princeton University Press, 1994.

Freud, S. (1887–1904) *The Complete Letters of Sigmund Freud to Wilhelm Fliess, 1887–1904*. Cambridge, Mass.: Harvard University Press, 1985.

Freud, S. (1895) 'Frau Emmy von N., age 40, from Livonia'. In: S. Freud and J. Breuer, *Studies in Hysteria*. Harmondsworth: Penguin Modern Classics, 2004, pp. 51–108.

Freud, S. (1899) 'Leonardo da Vinci and a memory of childhood'. In: S. Freud, *The Uncanny*. Harmondsworth: Penguin Books, 2003, pp. 45–120.

Freud, S. (1900) *The Interpretation of Dreams*. Harmondsworth: Volume 4, Penguin Freud Library. 1976.

Freud, S. (1905) 'Fragment of an analysis of hysteria (Dora)'. In: S. Freud, *The Psychology of Love*. Harmondsworth: Penguin Books, 2006, pp. 1–110.

Freud, S. (1909) 'Some remarks on a case of obsessive-compulsive neurosis [The "Ratman"]'. In: S. Freud, *The 'Wolfman' and Other Cases*. Harmondsworth: Penguin Books, 2002, pp. 123–202.

Freud, S. (1912) 'A note on the unconscious in psychoanalysis', In: S. Freud, *On Metapsychology: The Theory of Psychoanalysis*. Harmondsworth: Volume 11, Penguin Freud Library, 1984, pp. 50–57.

Freud, S. (1913) 'Totem and taboo: Some correspondences between the psychical live of savages and neurotics'. In: S. Freud, *On Murder, Mourning and Melancholia*. Harmondsorth: Penguin Books, 2005, pp. 5–166.

Freud, S. (1915) 'The Unconscious', In: S. Freud, *The Unconscious*. Harmondsworth: Penguin Books, 2005, pp. 47–86.

Freud, S. (1918) 'From the history of an infantile neurosis [The "Wolfman"]'. In S. Freud, *The 'Wolfman' and other cases* Harmondsworth: Penguin Books, 2002, pp. 203–320.

Freud, S. (1919) 'The uncanny'. In: S. Freud, *The Uncanny.* Harmondsworth: Penguin Books, 2003, pp. 123–162.

Freud, S. (1921) 'Psychoanalysis and telepathy'. In: *Psychoanalysis and the Occult,* 1953 (ed. G. Devereaux), 56–68. London: Souvenir Press edition, 1974.

Freud, S. (1922) 'Dreams and telepathy', In: *Psychoanalysis and the Occult,* 1953 (ed. G. Devereaux), 69–87. London: Souvenir Press edition, 1974.

Freud, S. (1923) 'The ego and the id'. In: S. Freud, *Beyond the Pleasure Principle and Other Writings.* Harmondsworth: Penguin Books, 2003, pp. 105–149.

Freud, S. (1925) 'The occult significance of dreams'. In: *Psychoanalysis and the Occult,* 1953 (ed. G. Devereaux), 87–90. London: Souvenir Press edition, 1974.

Freud, S. (1927) 'Fetishism'. In: S. Freud, *The Unconscious.* Harmondsworth: Penguin Books, 2005, pp. 93–100.

Freud, S. (1930) 'Civilization and its Discontents'. In: S. Freud, *Civilization and its Discontents.* Harmondsworth: Penguin Books, 2002, pp. 1–82.

Freud, S. (1933) 'Dreams and the occult'. In: *Psychoanalysis and the Occult,* 1953 (ed. G. Devereaux), 91–109. London: Souvenir Press edition, 1974.

Gallop, J. (1988) *Thinking Through the Body.* New York: Columbia University Press.

Ginsburg, E. (1996) 'The politics of passing'. In: *Passing and the Fictions of Identity* (ed. E. Ginsburg), 1–18. Durham, N.C.: Duke University Press.

Graham, S. (2010) *Cities under Siege: The New Military Urbanism.* London: Verso.

Gregg, M. and Seigworth, G. (eds.) (2010) *The Affect Theory Reader.* Durham, N.C.: Duke University Press.

Griffin, J. H. (1961) *Black Like Me.* Boston: Houghton, 1977.

Gurney, E., Myers, F. W. H. and Podmore, F. (1886) *Phantasms of the Living, Volume 1.* London: Rooms of the Society for Psychical Research and Trüber and Co.

Hamilton, T. (2009) *Immortal Longings: F. W. H. Myers and the Victorian Search for Life after Death.* Exeter: Imprint Academic.

Hannigan, J. (1998) *Fantasy City: Pleasure and Profit in the Postmodern Metropolis.* London: Routledge.

Haraway, D. (2008) *When Species Meet.* Minneapolis: University of Minnesota Press.

Harvey, D. (1993) 'Class relations, social justice and the politics of difference'. In: *Place and the Politics of Identity* (eds. M. Keith and S. Pile), 41–66. London: Routledge.

Harvey, D. (2013) *Rebel Cities: From the Right to the City to the Urban Revolution.* London: Verso.

Hewit, M. A. (2014) 'Freud and the psychoanalysis of telepathy: commentary on Claudie Massicotte's "Psychical Transmissions"', *Psychoanalytic Dialogues.* 24, pp. 103–108.

Highmore, B. (2011) *Ordinary Lives: Studies in the Everyday.* London: Routledge.

Hodkinson, S. (2018) 'Grenfell foretold: A very neoliberal tragedy'. In: *Social Policy Review 30: Analysis and Debate in Social Policy, 2018* (eds. C. Needham, E. Heins and J. Rees), 5–26. Bristol: Bristol University Press.

Hodkinson, S. (2019) *Safe as Houses: Private Greed, Political Negligence and Housing Policy after Grenfell.* Manchester: Manchester University Press.

Hoffman, M. (2000) 'The wolf and the seven little kids'. In: M. Hoffman, *The Macmillan Treasury of Nursery Stories.* Basingstoke: Macmillan Children's Books, pp. 40–44.

Holloway, J. (2006) 'Enchanted spaces: The séance, affect, and geographies of religion', *Annals of the Association of American Geographers.* 96(1), pp. 182–187.

Irigaray, L. (1977) *This Sex Which Is Not One*. Ithaca, New York: Cornell University Press, 1985.

James, L. (1995) *The Golden Warrior: The Life and Legend of Lawrence of Arabia. Revised and updated edition*. London: Abacus.

Johnson, A. (2020) 'Throwing our bodies against the white background of academia', *Area*. 52(1), pp. 89–96.

Johnson, R. (2020) *Lawrence of Arabia on War: The Campaign in the Desert 1916–1918*. London: Bloomsbury Publishing.

Jones, C. (1949) 'An end to the neglect of the problems of the negro woman!' *Political Affairs*. June, pp. 3–19.

Kanneh, K. (1998) *African Identities: Race, Nation and Culture in Ethnography, Pan-Africanisms and Black Literatures*. London: Routledge.

Kaplan, C. (ed.) (2007) *Passing: Nella Larsen*. New York: W. W. Norton and Co.

Kawash, S. (1996) 'The autobiography of an ex-colored man: (passing for) black passing for white'. In: *Passing and the Fictions of Identity* (ed. E. Ginsburg), 59–74. Durham, N.C.: Duke University Press.

Kingsbury, P. (2007) 'The extimacy of space', *Social and Cultural Geography*. 8(2), pp. 235–258.

Kingsbury, P. (2017) 'Uneasiness in culture, or negotiating the sublime distances towards the big Other', *Geography Compass online journal*. 11(6), e12316.

Kingsbury, P. and Pile, S. (2014) 'Introduction: The unconscious, transference, drives and repetition and other things tied to geography'. In: *Psychoanalytic Geographies*, 2014 (eds. P. Kingsbury and S. Pile), 1–38. Basingstoke: Ashgate.

Kivland, S. (1999) *A Case of Hysteria*. London: Book Works.

Kivland, S. (ed.) (2014) *Folles de Leur Corps/Crazy about their Bodies*. London: CG in association with Ma bibliothèque.

Kobayashi, A. and Peake, L. (1994) 'Unnatural discourses: "Race" and gender in geography', *Gender, Place and Culture*. 1(2), pp. 225–243.

Korda, M. (2010) *Hero: The Life and Legend of Lawrence of Arabia*. London: Aurum Press, 2012.

Kogan, I. (1988) 'The second skin', *International Review of Psycho-analysis*. 15, pp. 251–260.

Lacan, J. (1959) 'On a question preliminary to any possible treatment of psychosis'. In: J. Lacan, *Écrits: A Selection*. London: Tavistock, 1977, pp. 179–225.

Lacan, J. (1966) *Écrits*. New York: Norton, 2006.

Lacan, J. (1973) 'L'Étouridt [Stunned]' *Scilicet*. 4, pp. 5–52.

Laclos, P. (1783) *De L'Éducation des Femmes*. Paris: Jérôme Millon, 1991.

Lafrance, M. (2009) 'Skin and self: Cultural theory and Anglo-American psychoanalysis', *Body and Society*. 15(3), pp. 3–24.

Lamont, P. (2005) *The First Psychic: The Peculiar Mystery of a Notorious Victorian Wizard*. London: Abacus.

Lamont, P. (2013) *Extraordinary Beliefs: A Historical Approach to a Psychological Problem*. Cambridge: Cambridge University Press.

Larsen, N. (1928) *Quicksand*. London: Serpent's Tail, 1989.

Larsen, N. (1929) *Passing*. London: Serpent's Tail, 1989.

Lawrence, T. E. (1926) *Seven Pillars of Wisdom*. Harmondsworth: Penguin Books, 1935/1962.

Lazarre, J. (1997) *Beyond the Whiteness of Whiteness: Memoir of a White Mother of Black Sons.* Durham, N.C.: Duke University Press.

Le Bon, G. (1895) *The Crowd: A Study of the Popular Mind.* New York: Routledge, 1995.

Lefebvre, H. (1974) *The Production of Space.* Oxford: Basil Blackwell, 1991.

Locke, A. (1925a) 'The New Negro'. In: *The New Negro: Voices of the Harlem Renaissance* (ed. A. Locke), 3–16. New York: Touchstone.

Locke, A. (ed.) (1925b) *The New Negro: Voices of the Harlem Renaissance.* New York: Touchstone.

Longhurst, R. and Johnston, L. (2014) 'Bodies, gender, place and culture: 21 years on', *Gender, Place and Culture.* 21(3), pp. 267–278.

Longhurst, R. (2001a) *Bodies: Exploring Fluid Boundaries.* London: Routledge.

Longhurst, R. (2001b) 'Breaking corporeal boundaries: Geographies that matter: pregnant bodies in public spaces'. In: *Contested Bodies* (eds. J. Hussard and R. Holliday), 81–94. London: Routledge.

Lott, E. (1993) *Love and Theft: Blackface Minstrelsy and the American Working Class.* New York: Oxford University Press.

Luckhurst, R. (2002) *The Invention of Telepathy.* Oxford: Oxford University Press.

Macey, D. (2000) *Frantz Fanon: A Biography.* New York: Picador Press.

Macleod, G. (2018) 'The Grenfell Tower atrocity: Exposing urban worlds of inequality, injustice, and impaired democracy', *City.* 22(4), pp. 460–489.

Madden, D. J. (2017) 'Editorial: A catastrophic event', *City.* 21(1), pp. 1–5.

Mafe, D. (2008) 'Self-made women in a (racist) man's world: The "tragic" lives of Nella Larsen and Bessie Head,' *English Academy Review.* 25(1), pp. 66–76.

Mahony, P. (1996) *Freud's Dora: A Psychoanalytic, Historical and Textual Study.* New Haven: Yale University Press.

Mahtani, M. (2015) *Mixed Race Amnesia: Resisting the Romanticization of Multiraciality.* Toronto: UBC Press.

Mantegazza, P. (1896) *Physiology of Love and Other Writings.* Toronto: University of Toronto Press, 2007.

Martin, L. and Secor, A. (2013) 'Towards a post-mathematical topology', *Progress in Human Geography.* 38(3), pp. 420–438.

Marx, K. (1867) *Capital, Volume 1.* Harmondsworth: Penguin; *Le Capital* (French edition originally published 1875, Paris: Maurice Lachâtre et Cie).

Massicotte, C. (2014) 'Psychical transmissions: Freud, spiritualism, and the occult', *Psychoanalysis Dialogues.* 24, pp. 88–102.

Massumi, B. (2002) *Parables of the Virtual: Movement, Affect, Sensation.* Durham, N.C.: Duke University Press.

Mayer, E. L. (2001) 'On "telepathic dreams"?: An unpublished paper by Robert Stoller', *Journal of the American Psychoanalytic Association.* 49(2), pp. 629–665.

McCormack, D. (2006) 'For the love of pipes and cables: A response to Deborah Thien', *Area.* 38(3), pp. 330–332.

McCormack, D. (2008) 'Engineering affective atmospheres on the moving geographies of the 1897 Andrée Expedition' *Cultural Geographies.* 15, pp. 413–430.

McDowell, D. E. (1986) 'Introduction'. In: N. Larsen, *Quicksand* and *Passing.* New Brunswick: Rutgers University Press, pp. ix–xxxv.

McKittrick, K. (2000a) '"Black and 'Cause I'm Black I'm Blue": Transverse racial geographies in Toni Morrison's *The Bluest Eye*', *Gender, Place & Culture.* 7(2), pp. 125–142.

McKittrick, K. (2000b) '"Who do you talk to, when a body's in trouble?"': M. Nourbese Philip's (un)silencing of black bodies in the diaspora', *Social & Cultural Geography.* 1(2), pp. 223–236.

McKittrick, K. (2006) *Demonic Grounds: Black Women and the Cartographies of Struggle.* Minneapolis: University of Minnesota Press.

McKittrick, K. (2011) 'On plantations, prisons, and a black sense of place', *Social & Cultural Geography.* 12(8), pp. 947–963.

Merleau-Ponty, M. (1945) *The Phenomenology of Perception.* London: Routledge & Kegan Paul, 1962.

Miller, P. (2012) 'How Emmy silenced Freud into analytic listening'. In: *On Freud's 'On Beginning the Treatment'* (eds. G. Saragnano and C. Seulin), Chapter 16, 17 pages. London: Taylor and Francis, 2018.

Moi, T. (1981) 'Representation of patriarchy: Sexuality and epistemology in Freud's "Dora"', *Feminist Review.* 9 (Autumn), pp. 60–74.

Morehouse, D. (1996) *Psychic Warrior: True Story of the CIA's Paranormal Espionage Program.* New York: St Martin's Press.

Morrison, M.-B. (1964) *Jungle West 11.* London: Tandem Books.

Morrissey, K. (1995) *Dora – A case of hysteria.* London: Nick Hern Books.

Murji, K. and Solomos, J. (eds.) (2005) *Racialization: Studies in Theory and Practice.* Oxford: Oxford University Press.

Myers, F. W. H. (1903) *Human Personality and Its Survival of Bodily Death.* Charlottesville: Hampton Roads, 2001.

Nast, H. and Pile, S. (eds.) (1998) *Places through the Body.* London: Routledge.

Nayak, A. (2011) 'Geography, race and emotions: Social and cultural intersections', *Social & Cultural Geography.* 12(6), pp. 548–562.

Noxolo, P. (2018) 'Laughter and the politics of place-making'. In: *The Fire Now: Anti-racist Scholarship in Times of Explicit Racial Violence*, 2018 (eds. A. Johnson, R. Joseph-Salisbury and B. Kamunge), Chapter 22. London: Zed Books.

Obholzer, K. (1980) The Wolf-Man – Sixty Years Later: Conversations with Freud's Patient. London: Routledge & Kegan Paul, 1982.

Ogden, T. (1989) *The Primitive Edge of Experience.* New Brunswick: Jason Aronson.

Olund, E. (2009) '*Traffic in souls*: The "new woman", whiteness and mobile self-possession', *Cultural Geographies.* 16(4), pp. 485–504.

Pappenheim, E. (1980) 'Freud and Gilles de la Tourette: Diagnostic speculations on "Frau Emmy von N."', *The International Journal of Psychoanalysis.* 7, pp. 265–277.

Paterson, M. (2007) *The Senses of Touch: Haptics, Affects and Technologies.* Oxford: Berg.

Pavda, G. (2005) 'Dreamboys, meatmen and werewolves: Visualizing erotic identities in all-male comic strips', *Sexualities.* 8(5), pp. 587–599.

Pearsall, R. (1972) *The Table-Rappers: The Victorians and the Occult.* Stroud: Sutton Publishing, 2004.

Peter, J. D. (1999) *Speaking into the Air: A History of the Idea of Communication.* Chicago: Chicago University Press.

Peters, J. D. (2010) 'Broadcasting and schizophrenia', *Media, Culture and Society.* 32(1), pp. 123–140.

Phillips, M. and Phillips, T. (2009) *Windrush: The Irresistable Rise of Multi-racial Britain.* London: Harper Collins.

Pile, S. (1996) *The Body and the City: Psychoanalysis, Space and Subjectivity.* London: Routledge.

Pile, S. (1999) 'The heterogeneity of cities'. In: *Unruly Cities? Order/Disorder* (eds. S. Pile, C. Brook and G. Mooney), 7–42. London: Routledge.

Pile, S. (2000) 'The troubled spaces of Frantz Fanon'. In: *Thinking Space* (eds. M. Crang and N. Thrift), 260–277. London: Routledge.

Pile, S. (2005a) *Real Cities: Modernity, Space and the Phantasmagorias of City Life*. London: Sage.

Pile, S. (2005b) 'In the footsteps of angels: Tim Brennan's "Museum of Angels" guided walk', *Cultural Geographies*, 12(4), pp. 521–526.

Pile, S. (2009) 'Topographies of the body-and-mind: Skin ego, body ego, and the film *Memento*', *Subjectivity*. 27, pp. 134–154.

Pile, S. (2010a) 'Emotions and affect in recent human geography', *Transactions of the Institute of British Geographers*. 35(1), pp. 5–20.

Pile, S. (2010b) 'Intimate distance: The unconscious dimensions of the rapport between researcher and researched', *The Professional Geographer*. 62(4), pp. 483–495.

Pile, S. Bartolini, N. and MacKian, S. (2019) '"Creating a World for Spirit": Affectual infrastructures and the production of a place for affect', *Emotion, Space and Society*, 30, pp. 1–8.

Pile, S. and Keith, M. (eds.) (1997) *Geographies of Resistance*. London: Routledge.

Platt, C. B. (2010) 'The medium and the matrix: Unconscious information and the therapeutic dyad', *Journal of Consciousness Studies*. 16(9), pp. 55–76.

Powell, R. J. (1997) 'Re/Birth of a Nation'. In: *Rhapsodies in Black: art and the Harlem Renaissance* (ed. J. Skipworth), 14–33. London and Berkeley: The Hayward Gallery, The Institute of International Visual Arts and the University of California Press.

Price, P. (2012) 'Race and ethnicity II: Skin and other intimacies', *Progress in Human Geography*. 37(4), pp. 578–586.

Prosser, J. (1998) *Second Skins: The Body Narratives of Transsexuality*. New York: Columbia University Press.

Radical Housing Network, Hudson, B. and Tucker, P. (2019). 'Struggles for Social Housing Justice'. In *After Grenfell: Violence, Resistance and Response*, 2019 (eds. D. Bulley, J. Edkins and N. El-Enany), 62–74. London: Pluto Press.

Rancière, J. (2001) *The Aesthetic Unconscious*. Cambridge: Polity Press, 2009.

Rancière, J. (2004) *The Politics of Aesthetics*. London: Bloomsbury Press.

Rancière, J. (2010) *Dissensus: On Politics and Aesthetics*. London: Bloomsbury Press.

Rancière, J. (2012) *Proletarian Nights: The Workers' Dream in Nineteenth-century France*. London: Verso.

Redman, P. (2009) 'Affect revisited: Transference-countertransference and the unconscious dimensions of affective, felt and emotional experience', *Subjectivity*. 26(1), pp. 51–68.

Renwick, D. (2019) 'Organising on mute'. In: *After Grenfell: Violence, Resistance and Response*, 2019 (eds. D. Bulley, J. Edkins and N. El-Enany), 19–46. London: Pluto Press.

Reynolds, D. (2009) 'Response to "skin and the self": Cultural theory and Anglo-American psychoanalysis', *Body and Society*. 15(3), pp. 25–32.

Rockhill, G. (2004) 'Appendix I'. In: *The Politics of Aesthetics* (J. Rancière), 83–98. London: Bloomsbury Press.

Roof, J. (1989) 'The match in the crocus: Representations of lesbian sexuality'. In: *Discontented Discourses: Feminism/Textual Intervention/Psychoanalysis* (eds. M. S. Barr and R. Feldstein), 100–116. Urbana and Chicago: University of Illinois Press.

Routledge, P. (2012) 'Sensuous solidarities: emotion, politics and performance in the Clandestine Rebel Clown Army', *Antipode*. 44(2), pp. 428–452.

Ruddick, S. (2010) 'The politics of affect: Spinoza in the work of Negri and Deleuze', *Theory, Culture and Society*. 27(4), pp. 21–45.

Rustin, M. (2009) 'Esther Bick's legacy of infant observation at the Tavistock – Some reflections 60 years on', *Infant Observation*. 12(1), pp. 29–41.

Saad, T. and Carter, P. (2005) 'The entwined spaces of "race", sex and gender', *Gender, Place & Culture*. 12(1), pp. 49–51.

Said, E. (1978) *Orientalism*. Hardmondsworth: Penguin Books.

Saldanha, A. (2006) 'Reontologising race: The machine geography of phenotype', *Environment and Planning D: Society and Space*. 24, pp. 9–24.

Saldanha, A. (2010) 'Skin, affect, aggregation: Guattarian variations on Fanon', *Environment and Planning A*. 42, pp. 2410–2427.

Sánchez, M. C. and Schlossberg, L. (eds.) (2001) *Passing: Identity and Interpretation in Sexuality, Race, and Religion*. New York: New York University Press.

Sborgi, A. V. (2019) 'Grenfell on screen'. In: *After Grenfell: Violence, Resistance and Response*, 2019 (eds. D. Bulley, J. Edkins and N. El-Enany), 97–118. London: Pluto Press.

Shabazz, R. (2015) *Spatializing Blackness: Architectures of Confinement and Black Masculinity in Chicago*. Chicago: University of Illinois Press.

Shamdasani, S. (1994) 'Encountering Hélène: Théodore Flournoy and the genesis of subliminal psychology'. In: T. Flournoy, *From India to the Planet Mars: A Case of Multiple Personality with Imaginary Languages*. Princeton: Princeton University Press, pp. xi–li.

Sharp, J., Routledge, P., Philo, C. and Paddison, R. (eds.) (2000) *Entanglements of Power: Geographies of Domination/Resistance*. London: Routledge.

Shildrick, T. (2018) 'Lessons from Grenfell: Poverty propaganda, stigma and class power', *The Sociological Review*. 66(4), pp. 783–798.

Silverman, K. (1992) *Male Subjectivity at the Margins*. London: Routledge.

Simonsen, K. (2005) 'Bodies, sensations, space and time: The contribution of Henri Lefebvre', *Geografiska Annaler B: Human Geography*. 87(1), pp. 1–14.

Simonsen, K. (2012) 'In quest of a new humanism: embodiment, experience and phenomenology as critical geography', *Progress in Human Geography*. 37(1), pp. 10–26.

Simonsen, K. and Koefoed, L. (2020) *Geographies of Embodiment: Critical Phenomenology and the World of Strangers*. London: Sage.

Simpson, C. and Knightley, P. (1969) *The Secret Lives of Lawrence of Arabia*. London: Panther Books. (Correcting and augmenting four articles that appeared in the *Sunday Times* on 9th, 16th, 23rd, and 30th of June.)

Sinclair, U. (1930) *Mental Radio*. Newbury Port, Mass.: Hampton Roads, 2001.

Skipworth, J. (ed.) (1997) *Rhapsodies in Black: Art and the Harlem Renaissance*. London and Berkeley: The Hayward Gallery, The Institute of International Visual Arts and the University of California Press.

Smith, P. H. (2005) *Reading the Enemy's Mind: Inside Star Gate – America's Psychic Espionage Program*. New York: Forge.

Stark, S. (1996) *Female Tars: Women Aboard Ship in the Age of Sail*. Annapolis, Maryland: Naval Institute Press.

Steele, V. (2014) 'Introduction'. In: *Exposed: A History of Lingerie* (ed. C. Hill), 7–15. New Haven: Yale University Press.

Stewart, K. (2007) *Ordinary Affects*. Durham, N.C.: Duke University Press.

Stoler, A. (2002) *Carnal Knowledge and Imperial Power: Race and the Intimate in Colonial Rule*. Berkeley: University of California Press.

Stoller, R. (1973) 'Telepathic Dreams?' In: E. L. Mayer, 'On "Telepathic Dreams?" An unpublished paper by Robert J. Stoller', 2001, *Journal of the American Psychoanalytic Association*. 49(2), pp. 635–650.

Straughan, E. (2014) 'The uncanny in the beauty salon'. In: *Psychoanalytic Geographies* (eds. P. Kingsbury and S. Pile), 295–306. London: Routledge, 2016.

Straughan, E. (2015) 'Entangled corporeality: Taxidermy practice and the vibrancy of dead matter', *GeoHumanities*. 1(2), pp. 363–377.

Sullivan, S. (2004) 'Ethical slippages, shattered horizons, and the zebra striping of the unconscious: Fanon on social, bodily, and psychical space', *Philosophy and Geography*. 7(1), pp. 9–24.

Sulloway, F. (1979) *Freud, Biologist of the Mind: Beyond the Psychoanalytic Legend*. Cambridge, Mass.: Harvard University Press, 1992.

Szczelkun, S. and Iles, A. (eds.) (2012) *Agit-Disco*. Berlin: Mute Books.

Tate, C. (1980) 'Nella Larsen's *Passing*: A problem of interpretation', *Black American Literature Forum*. 14 (Winter) pp. 180–246.

Tate, C. (1992) *Domestic Allegories of Political Desire: The Black Heroine's Text at the Turn of the Century*. Oxford: Oxford University Press.

Taylor, M. (2002) *Harlem: Between Heaven and Hell*. Minneapolis: University of Minnesota Press.

Thomas, L. (1924) *With Lawrence in Arabia*. London: Prion, 2002.

Thrift, N. (2004) 'Intensities of feeling: Towards a spatial politics of affect', *Geografiska Annaler*. 86B, pp. 57–78.

Thrift, N. (2008) *Non-representational Theory: Space | Politics | Affect*. London: Routledge.

Todorov, T. (1982) *Theories of the Symbol*. Oxford: Basil Blackwell.

Tögel, C. (1999) '"My bad diagnostic error": Once more about Freud and Emmy v. N. (Fanny Moser)', *The International Journal of Psychoanalysis*. 80(6), pp. 1165–1173.

Tolia-Kelly, D. P. (2016) 'Feeling and being at the (postcolonial) museum: Presencing the affective politics of "race" and culture', *Sociology*. 50(5), pp. 896–912.

Tolia-Kelly, D. P. (2019) 'Rancière and the re-distribution of the sensible: The artist Rosanna Raymond, dissensus and postcolonial sensibilities within the spaces of the museum', *Progress in Human Geography*. 31(1), pp. 123–140.

Ulnik, J. (2007) *Skin in Psychoanalysis*. London: Karnac Books.

Valentine, G. (2007) 'Theorizing and researching intersectionality: A challenge for feminist geography', *The Professional Geographer*. 59(1), pp. 10–21.

Venturi., R., Brown, D. S. and Izenour, S. (1972) *Learning from Las Vegas: The Forgotten Symbolism of Architectural Form* (revised edition 1977). Boston: MIT Press.

Wald, G. (2000) *Crossing the Line: Racial Passing in Twentieth-Century U.S. Literature*. Durham, N.C.: Duke University Press.

Wegener, M. (2016) 'Psychoanalysis and topology – Four vignettes'. In: *Psychoanalysis: Topological Perspectives* (eds. M. Friedman and S. Tomšic), 31–52. Bielefeld: transcript Verlag.

Wetherell, M. (2012) *Affect and Emotion: A New Social Science Understanding*. London: Sage.

Wilson, J. (1989) *Lawrence of Arabia: The Authorized Biography of T. E. Lawrence.* London: Heinemann.

Wooffitt, R. (2017a) 'Relational psychoanalysis and anomalous communication: continuities and discontinuities in psychoanalysis and telepathy', *History of the Human Sciences.* 30(1), pp. 118–137.

Wooffitt, R. (2017b) 'Shared subjectivities: Enigmatic and mundane intimacies', *Subjectivity.* 11(1), pp. 40–56.

Yuknavitch, L. (2012) *Dora: A Headcase.* Portland: Hawthorne Books.

Index

Please note that page references to Figures will be followed by the letter 'f'

Bodies, Affects, Politics: The Clash of Bodily Regimes, First Edition. Steve Pile.
© 2021 Royal Geographical Society (with the Institute of British Geographers).
Published 2021 by John Wiley & Sons Ltd.